Physical Asset Management

AN ORGANIZATIONAL CHALLENGE

by Dharmen Dhaliah

 FriesenPress

Suite 300 - 990 Fort St
Victoria, BC, V8V 3K2
Canada

www.friesenpress.com

Copyright © 2016 by Dharmen Dhaliah
First Edition — 2016

All rights reserved. No part of this book may be used or reproduced by any means, graphic, electronic, or mechanical, including photocopying, recording, taping or by any information storage retrieval system without the written permission of the copyright owner, except in the case of brief quotations embodied in critical articles and reviews.

The views and opinions expressed in this book are the author's personal thoughts. They are not intended to be a definitive set of instructions. Names, characters, businesses, places, events and incidents are either the products of the author's imagination or used in a fictitious manner. Any similarity or resemblance to actual persons, living or dead, or actual situations, cases or firms is purely coincidental.

ISBN
978-1-4602-8825-2 (Hardcover)
978-1-4602-8826-9 (Paperback)
978-1-4602-8827-6 (eBook)

1. Business & Economics, Infrastructure

Distributed to the trade by The Ingram Book Company

This book is dedicated to

*my late mother Pouspah and my late father Narainsamy,
who provided life*

to my wife Mitrini, who shares it

to my sons Yuvneesh and Yaneesh, who give it purpose

CONTENTS

Preface vii

Acknowledgements 1

Introduction 2

Chapter 1 — *Organization Paradigm Shifts* 8

Chapter 2 — *Typical Functional Areas of Organizations* 21

Chapter 3 — *Physical Asset Management's Fit in Organizations* . . . 34

Chapter 4 — *Challenges of Physical Asset Management* 53

Chapter 5 — *Operations* 75

Chapter 6 — *Maintenance Management* 85

Chapter 7 — *Reliability Engineering* 97

Chapter 8 — *Project Management* 108

Chapter 9 — *Procurement Management* 121

Chapter 10 — *Financial Planning* 130

Chapter 11 — *Physical Asset Management* 142

Chapter 12 — *Implementing Physical Asset Management* 159

Conclusion 181

Bibliography 184

Index 191

PREFACE

Asset management has been practiced for many years in different industries and business sectors. With time, it has developed into a new discipline and is now evolving into an overarching principle, in great demand both in the public and private sectors. The key factor behind this evolution is mainly due to our rapidly deteriorating assets/infrastructure (physical assets), in desperate need of more care and attention to continue to safely deliver essential services, while meeting ever-expanding needs and supporting expected rates of growth.

It should be pointed out early on that organizations have different types of assets such as tangible, intangible, human, financial and information assets. This book deals with management of tangible physical assets only such as equipment and infrastructure. Later in the book we will clarify the terms "asset" and "physical asset" in more details, but hereon we will use the term "physical asset management", which is the focus of this book..

Good physical asset management practices are indeed in place in many organizations, but they are either scattered within the organizations or imbedded in different functional areas. These functional areas manage distinct portions of a physical asset's lifecycle, forming functional silos which are sometimes in conflict or completely misaligned with each other. The bureaucratic silos, as they are often called make it challenging for organizations to manage their physical assets in an effective manner to realize maximum value, while achieving the right balance between providing the required levels of service, optimizing costs, and managing risks. The organizational silos may sometimes overlap with each other, causing duplication of activities, while others may be far apart, creating gaps within the physical assets lifecycle. These gaps and overlaps are detrimental to effective whole lifecycle physical asset/infrastructure management. With the advent of physical asset management standards, such as Publicly Available Specification (PAS) 55, ISO 55000, enforcement of acts and regulations from different level of governments, and stringent requirements from industry legislations, organizations in the public and private sectors are focused more than ever on adopting a holistic approach to managing their physical assets to remain in compliance, to obtain required funding at the right time, and to become more competitive.

For organizations to become more effective in holistic physical asset management, they have to review their strategic plan and vision, organizational structure, and business processes to streamline all activities and create maximum value from their physical assets. Implementing PAS55 or ISO55000 is an excellent way to start the transformation process, but it is very critical to determine how the holistic physical asset management approach will fit within organizations and how it can fulfill its mandate while collaborating with other key functional areas.

A few years ago, I attended a maintenance and physical asset management conference regrouping professionals and practitioners from different parts of the world. During the expert panel discussion, I asked the following question: "How does the physical asset management function fit in an organization and what are its relationships with the existing functional areas?"

The panelists, who are eminent global leaders in the field of physical asset management, tried to answer the question the best they could, but at that time, I was left with a deep sense of incompleteness. The message I got from the panel was that this burning question has been around for many years and there is no straight or direct answer. Over time, organizations have to figure it out and deal with it themselves.

In this book I will try to answer the above question and explore the intricacies and pitfalls when implementing physical asset management in organizations. The book will shed some light on how physical asset management fits in an organization and the challenges it encounters with the organizational culture and other functional areas.

During my twenty-six years' experience in the private and public sectors, I have had the opportunity to hold different roles in different functional areas such as operations, maintenance, reliability, project management and physical asset management. I was able to experience firsthand the challenges involved with managing physical assets from different angles and perspectives both at the tactical and at the strategic levels. This book is based on my experience, my observations and my personal views on how to approach the subject.

It is written in a plain and simple language without getting into the technical details of physical asset management to make it an easy read for everyone at any level of an organization. The role of the book is to promote awareness of what options maybe out there rather than a definitive set of instructions. Readers with different exposure and experience may not agree or be in approval with some of the contents. I will let the readers make their own judgement and figure out what would work for them.

The book also contains some interesting cases studies that replicate real life physical asset management challenges in the industry. This makes it a useful resource for use in physical asset management courses. You will notice that the case studies do not include solutions. This is done on purpose to allow readers to trigger discussions around the topic and come up with the right approach based on their own experience and operational context. The key here is to highlight and focus on those specific challenges – that's half the battle.

This book is dedicated to all physical asset management practitioners and professionals, to everyone dealing with physical assets in one way or the other, and to all decision-makers and leaders of organizations who recognize the need for a holistic physical asset management approach. Readers will find the book beneficial when they implement physical asset management in their organizations and it will help them be better prepared to face the challenges it entails.

ACKNOWLEDGEMENTS

This book would have never been possible without the inspiration of the many professionals, practitioners, and colleagues who I had the opportunity to know and work with throughout my whole career. Similar to me, they have gone through the same challenges while embarking on the journey to implement a holistic physical asset management approach in their organizations. I would like to extend my gratitude to all peers, colleagues, and friends who have encouraged me all the way to write this book.

A big thank you to Cindy Snedden and Cliff Williams for taking time off their busy schedule to review the book in its entirety and to provide their expert advices and recommendations.

I would also like to thank Maria Mark and Balraj Appadu who did a great job helping me with the illustrations and graphics, and Elvin Sunnassee for his invaluable comments and feedback.

Last but not least, I would like to thank my wonderful wife Mitrini and my two fabulous children Yuvneesh and Yaneesh, who have been patient and very supportive while I was busy writing the book.

INTRODUCTION

I-35W bridge collapses over Mississippi River; Elliot Lake mall roof collapses; Travel chaos for London commuters due to Underground train derailment; Oil rig explosion in Gulf of Mexico; Power failures hit millions in India; Commuter train derails in Granville, Flight 8641 crashed near Mozyr; The Camara dam failure causes tragic deluge. These are just a few of the unfortunately very long list of headlines that we are familiar with. The effects and consequences of those catastrophic incidents that made headlines due to physical asset/infrastructure failures are very significant – loss of lives, major impact to the environment, disruption to essential services, huge financial losses. All these sound quite scary, don't they? When you board that flight, do you ask yourself if it will reach its destination safely? When you cross that bridge on your way to work, do you feel confident you will reach the other end? Are we at risk when consuming any processed food product? How safe you and your family feel when you are enjoying one of those attraction rides while on vacation? How reliable are the railway crossing signals and lights? Most, if not all of us depend on physical assets/infrastructure on a daily basis to produce safe goods for consumption and to deliver essential services such as transportation, power, water, and communication. Goods can be anything tangible ranging from food, clothing, to cars, appliances, aircrafts, etc. Effective modes of transport includes high-quality roads, railroads, ports, and air transport, efficient electricity supply means free from interruptions and shortages, efficient water infrastructure provides safe and uninterrupted supply of drinking water, and finally, a solid and extensive telecommunications network provides rapid and free flow of information. Extensive and efficient physical asset/infrastructure is critical for ensuring the effective functioning of the economy. Whole nations depend on them to create a strong and competitive economy and support social development. However the situation is not quite bright. On one side you have the ageing physical assets and infrastructure in need of major investment in the form of rehabilitation or expansion and on the other side the limited amount of funding available. As pointed out in the McKinsey Global Institute (MGI) report *Bridging Global Infrastructure Gaps, June 2016*, "the world spends some $2.5 trillion a year on the transportation, power, water, and telecom systems that underpin economic activity and provide essential services. But this has not been enough to avoid significant gaps, and investment needs are only growing steeper." The report also estimates that the investment needs to average $3.3 trillion annually through 2030 just to support current economic growth projections, an equivalent of 3.8 percent of global GDP. Too many countries have been underinvesting for decades, which have not helped in bridging the global infrastructure gaps. For the last few years we have seen many of those countries reacting and trying to address the issue. In Canada for example the government has pledged to double spending in infrastructure to jump-start economic growth and address the infrastructure deficit gap. Similarly in the United States, where they recognize that infrastructure systems are failing to keep pace with the current and expanding needs, and investment in infrastructure is faltering, the government is pushing for funding as part of the Build America Investment Initiative. Other countries are taking similar actions to address their infrastructure needs and investment shortfalls via political and public discussions on the issue. However, there is a fundamental question: Is funding the only global infrastructure issue? The answer is clearly no. Funding strategies alone will not address the infrastructure deficit. The MGI's 2013 report *Infrastructure productivity: How to save $1 trillion a year*, demonstrates that improving project selection, delivery, and management of existing physical assets could translate into 40 percent savings. Basically the report says that we need to get more, better-quality infrastructure for less when selecting, designing and delivering infrastructure projects, and we need to make more out of the infrastructure already in place. As MGI's report states – "scaling up best practice could save an average of $1 trillion a year in infrastructure costs over the next 18 years." This boost in infrastructure productivity can only help to bridge the infrastructure gap faster! The MGI report further recommends three main levers to roll out proven best practice in organizations in order to improve infrastructure governance systems and productivity:

1. Improving project selection and optimizing infrastructure portfolios
2. Streamlining delivery
3. Making the most of existing infrastructure assets

The above three main levers sum up how physical assets/infrastructure needs to be managed from cradle to grave in a holistic manner. It must be recognized that the three levers equally apply to physical assets/infrastructure in private organizations as well. As mentioned earlier countries, governments at all levels and private businesses to that effect are doing their part by injecting billions of dollars in physical asset/infrastructure to close the investment gaps and continue to deliver essential services and required goods. But what are organizations in general doing about it? How ready are the organizations and how well-prepared are they to make the most of the available funds?

This book is for organizations facing those questions, for organizations dealing mainly with tangible physical assets such as buildings, infrastructure, equipment and different other types of machinery to produce goods or deliver services. In the book we will use "physical assets" to cover the whole spectrum of tangible assets organizations manage and the term "physical asset management" for the holistic approach of managing those assets. Chapter 2 elaborate further on these topics.

Over the last few years in various industries, both in the public and private sectors holistic physical asset management has been and is still being vastly discussed. And for good reason. First, and foremost, because of the increased interest in the condition of our ageing physical assets, and second because of the publication of the ISO 55000 suite of standards on physical asset management. With ISO 55000, we have seen the emergence of different companies, organizations, consulting firms, and conference organizers setting the stage to develop, discuss, and share physical asset management best practices and philosophies among practitioners. There has been a corresponding increase in the development and introduction of educational and certification courses in physical asset management.

We hear from the various industries, and their leaders, that there is a pressing need for a shift from the traditional physical asset maintenance and reliability function to a more holistic physical asset management approach; however, this mindset shift has brought confusion into the world of maintenance and reliability practitioners. Questions crop up:

- What is holistic physical asset management?
- Is it not the same as traditional physical asset maintenance?
- Who does or will do what?
- What skill sets are required to practice holistic physical asset management?

Many organizations, with good intentions to embrace physical asset management best practices, are struggling for answers. This book explores some of the questions about physical asset management function within an organization and attempts to clarify the confusions that currently exist. It also provides recommendations on how to overcome those confusions together with the prevailing organizational challenges.

There are very useful resources out there that can help and guide us in our quest to embrace holistic physical asset management and explore the best governance systems. Various standards, documents, papers, articles, conceptual models, and principles are available. All these resources serve mostly as guidelines and strategic directions. There are also several certifications, assessments, and educational programs that can enhance our knowledge in the field of physical asset management. And we should not forget about all the great textbooks, seminars, conferences, and online discussion forums that help share experiences and ideas.

But there is still a very key question that needs to be answered: Where and how do we start to implement physical asset management? To try to answer, or at least bring some clarity to this question, we need to explore and understand the following:

- What are the functions dealing with physical assets in a traditional asset-centric organization?
- What is physical asset management, itself, and what is its current role in different organizations?
- How does the physical asset management function fit into organizations and what challenges does it face?
- What are the roles of an Asset Manager and what competencies and skill sets are required for a physical asset management function to succeed?

Let us start off with this question: "How does physical asset management fit into your organization?"

This question has been bugging me for the last few years. I asked around, but could never get a clear answer, which was one of the main reasons that prompted me to write this book. This book will not give a definite answer on how organizations should structure themselves to practice physical asset management because there are so many other factors to consider from an organizational perspective. This book will, however, explore the intricacies and pitfalls of fitting physical asset management in the workplace, and will help organizations to better understand and plan for a successful physical asset management implementation.

First and foremost, it should be pointed out and made clear that physical asset management is not a new field. Many organizations have been practicing it for years, with varying level of success. Physical asset management practices are often "hidden" or executed within other functions or dispersed throughout several functional areas of the organization. Physical asset management could exist either as a separate function, as part of maintenance management, or even be integrated into the asset planning finance/accounting function. There is also a substantial difference in how physical asset management is practiced in the public and private sectors. Organizations in both sectors have been successful in their physical asset management journey in their own ways—the public sector more so from a planning perspective (top down) and the private sector from a lifecycle delivery activities perspective (bottom up). Both sectors, though, have faced challenges in closing important gaps to achieve excellence in physical asset management. In this book, we will discover and explore those gaps.

With the popularity of PAS 55, ISO 55000, other applicable standards and publications, organizations are now striving to either implement holistic physical asset management (if it does not exist) or to develop a governance framework around their physical asset management approach. It is not an easy task in itself, since it involves synergies between many functional areas in an organization and, most probably, radical changes in the ways organizations currently manage their business. But it is a worthy undertaking to realize maximum value from existing capital stock and to make the most of current and future investments.

CHAPTER 1 —
Organization Paradigm Shifts

"Never before in history has innovation offered promise of so much to so many in so short a time."
~Bill Gates

Organization transformation

Over the years, organizations have realized that only by constantly evolving and improving will they be able to meet their customers' new demands and higher expectations. They have learned how to make the leap from good to great organization, from good to great processes, from good to great results, and from good to great performance.

As Jim Collins states in his book, *Good to Great,* the transformation from good to great is a process comprised of three key elements: disciplined people, disciplined thought, and disciplined action. Organizations go through complete re-engineering by reviewing the way they do business at all different levels, by looking closely at their strengths and weaknesses, and by making changes for the betterment of the organization. This transformation is critical, in today's tough and ever-changing business environment, so as to remain competitive in the global marketplace. Trends such as globalization, supply chain management, e-commerce, and technology have a growing impact on organizations' strategies and operations.

Organizations have no choice but to embrace new ways of doing business so they can survive the changing business landscape. The transformation process is a normal occurrence in organizations, and it brings with it paradigm shifts, new ways of looking at things, and new ways of doing things.

The Evolution of Operations

If we go back in time to take a closer look at operations in organizations, we can appreciate how much operations have evolved over the years. Operations have evolved through different eras—the craft production era, the industrial revolution era, the lean production era, and now the data management era.

Figure 1.1 below illustrates those evolution eras together with the key business trendsetters that created the paradigm shifts from cost-oriented to value-oriented to service-oriented and finally to data-oriented.

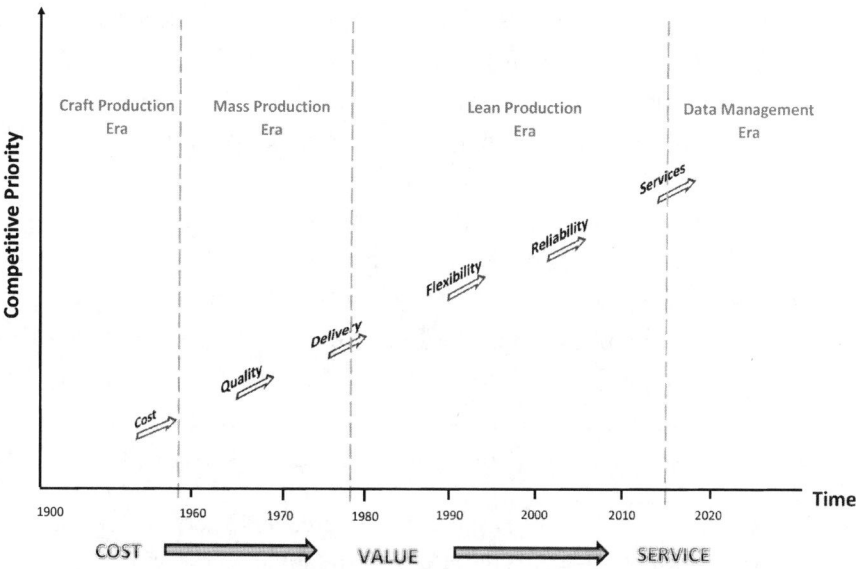

Fig. 1.1 The Evolution of Operations

Humans have been known for developing and using their skills to produce goods and provide services by applying simple and flexible tools. In the early years most of those goods and services were produced according to customer specifications and required enhanced craftsmanship. This era is known as the craft production era, or cottage industry, where production was slow and costly and work was carried out in the home by craftsmen or apprentices, who used handmade tools and equipment.

The late 1700s saw the birth of the factory system, where all workers came to work at a central location. This system brought in innovations that changed production, using machine power instead of human power. Large and expensive machines, such as the steam engine, were invented and completely changed the way goods were produced. This era was called the Industrial Revolution. Production of goods moved to a larger scale, and concepts such as division of labour and interchangeable parts were developed to support the mass production process. Scientific management emerged in the 1900s and brought widespread changes to the management of factories. Concepts of analysis, measurement of the technical aspects of work design, development of moving assembly lines, and mass production were introduced. This was led by the "father of scientific management," Frederick Taylor, whose methods emphasized maximizing outputs. A number of other pioneers also contributed hugely to scientific management principles, namely Frank Gilbreth, Lillian Gilbreth, and Henry Gantt. During that time, we also saw Henry Ford introduce the moving assembly line, which completely revolutionized the automotive industry. Between the 1930s and the 1970s, the Industrial Revolution's wave of change brought other changes to the operations landscape, such as the human relations movement, the computer age, and environmental issues and awareness.

During the 1980s, business operations entered a new era. This paradigm shift was influenced by Japanese manufacturers, who developed and refined management practices to increase the productivity of their operations and the quality of their products. The era saw the introduction of revolutionary philosophies and strategies such as total quality management, just-in-time, business process reengineering, flexibility, supply chain management, and e-commerce. This era, up to just recently, is known as the lean production era, where the focus is on elimination of waste and continuous improvement.

Now we are in the fourth era—a data revolution era—where data is emerging as a new asset class that will have to be handled. Mobile communications, social media, and sensors are blurring the boundaries between people, the Internet, and the physical world. Data is building up increasingly on who we are, who we know, where we are, where we have been, and where we plan to go. Data storage, safeguards, mining, and analyzing will be our next preoccupation, if it is not already.

The fact is that operations' evolution over the years has been the catalyst for paradigm shifts in other areas of organizations such as quality, technology, maintenance, and information systems, just to name a few.

The Evolution of Quality

Over the years, quality management processes have evolved to meet operations' different trendsetting expectations. The concept of quality has existed for many years—it is only the dimensions of quality that have changed and evolved over time.

During the cottage industry era, the business owner, product designer, craftsman, and salesman are often the same person in a craft business. His or her customers are friends and neighbours. The owner/operator of a craft business experiences the customer's satisfaction or dissatisfaction very directly and this drives a keen interest in quality management often described as "pride". This would define their perception of quality. Besides the owner's good reputation, there were no standards against which to compare, and every product was different in shape, size, or colour. During the Industrial Revolution, with the advent of mass production, quality had to be managed differently. This meant the introduction of product inspections— entailing sampling techniques and statistical analysis—to ensure products met specifications. With the great contribution of some prominent quality gurus—namely Walter A. Shewhart, W. Edwards Deming, Joseph M. Juran, Armand V. Feigenbaum, Philip B. Crosby, and Kaoru Ishikawa—the concept of quality took on a broader meaning and has evolved over time as shown in Fig. 1.2 below.

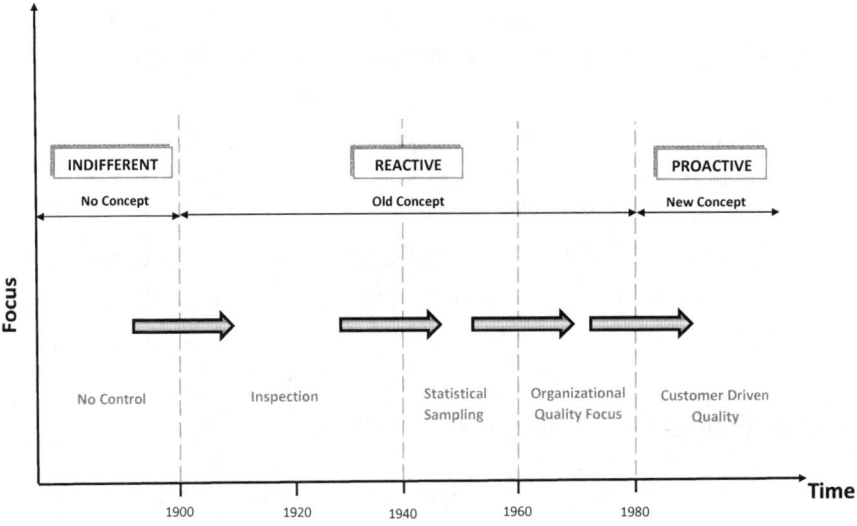

Fig. 1.2 The Evolution of Quality

Quality began to be viewed not only as a function of the operation process, but also as a function that needed to be managed by the whole organizations. During the twentieth century, companies started to take action to move from the reactive mode of quality inspection and correction to a more proactive approach of building quality in the product and process design. This brought significant changes to dimensions and focus of quality management, and to the global market. To be able to survive, organizations had to undertake major changes in the way they approached quality management, so they started investing in quality management programs, such as ISO 9000, Quality Function Design, Six Sigma, or the Toyota Production System. In many industries, quality excellence became a standard for doing business, as demonstrated by the different quality awards and certifications that are coveted by organizations. Successful companies understood that excellence in producing quality goods and delivering quality services provides a competitive edge. They started to develop strategies such as Total Quality Management (TQM), which sweeps across the entire organization to put customers first and defines quality as meeting or exceeding customers' expectations. Nowadays a lot of emphasis is put on data quality which will help make better forecast and hence better decisions.

The Evolution of Technology

Another aspect of operations management that has seen lots of changes over the years is technology. Since the beginning of the lean production era, there has been unprecedented growth in technological advances. There are three primary types of technologies: product technology, process technology, and information technology. Notable innovators, great minds in great organizations, have made their marks-from Alexander Graham Bell to Steve Jobs, from the Ford Motor Company to Tesla Motors-with amazing product innovations from the IBM mainframe to the sleek Microsoft Surface tablet.

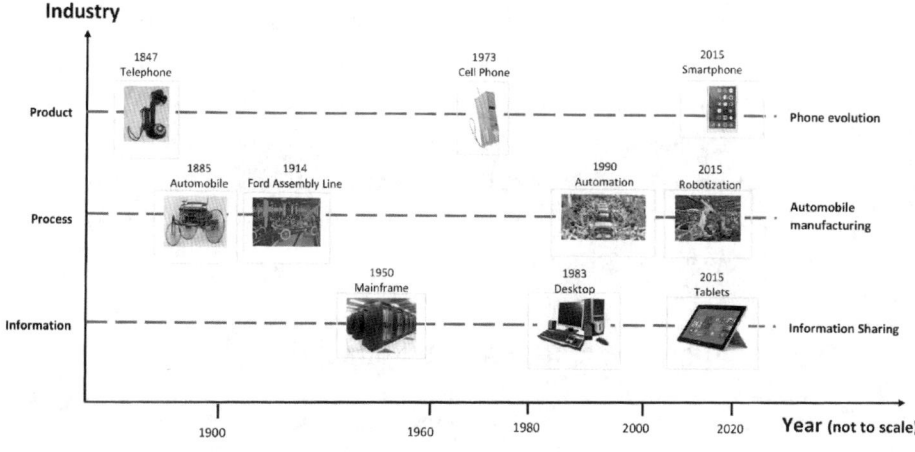

Fig. 1.3 The Evolution of Technology

All three types of technologies have undergone massive evolution as shown in Fig. 1.3 above and have revolutionized the way organizations operate. This evolution can be seen from two different angles.

The *push angle* is evidenced in organizations that continuously develop new technologies, and which innovate to improve productivity and profitability, and gain a competitive advantage on their competitors. An example of that is the evolution from early bag phones to the iPhone.

The *pull angle* is where customers dictate new needs and demands or when regulators inform companies of upcoming changes to regulations. This angle forces organizations to innovate or develop new technologies to meet those higher expectations. For example, in the auto industry, a change in customers' needs and in regulations requiring hands-free phone use has forced auto manufacturers to develop and introduce new technologies.

Obviously, the evolution of technology has brought significant benefits to our way of life, and made many organizations prosper. Embracing new technologies can help organizations gain competitive advantage, for example, by improving the speed and quality of processes, by designing products in innovative ways, or by enabling companies to share real-time information across the globe. But, at the same time, technological advancement has put a huge burden on how we manage operations. Everything comes with a higher price tag—cost to operate, cost to repair, obsolescence, financial justification—that needs to be assessed. To survive in this maze of fast technological advancement—be it from a product, process, or information perspective—organizations have to be very alert so they can keep abreast of changes to quickly evaluate the situation, and make the right business decisions.

The Evolution of Maintenance & Reliability

When we look back at how technology has evolved over the years, we can certainly understand how equipment and physical assets associated with those technologies have evolved as well. Not only has the equipment become more complex and more refined, but also expectations on its performance and output have increased. From a business point of view, investing in new equipment or physical assets to support a new technology implies improving output and hence improving return on investment. There is no doubt that the maintenance function must have evolved over time to be able to keep pace with the changing technological landscape and complexity of physical assets.

During the cottage industry and the birth of the factory system, there was not much focus on maintenance, and at that time machinery & equipment were not too complex. Things would get fixed when they broke, and the repercussions for breakage and downtime were not serious. However, during the Industrial Revolution, where the focus of operations shifted to mass production, volume and consistent quality of products became crucial. Equipment became more mechanized and complex, as demonstrated by the introduction of the assembly line. Equipment downtime and throughput of the production line became vital elements of operations. For example, with the application of the scientific management approach, neither Frederick Taylor nor Henry Ford would have been happy to see an assembly line stopped for a long time due to an equipment breakdown. On top of that, during World War II, the pressure from increased demand for reliable goods made industries more dependent on machinery and equipment. The maintenance function started to evolve into preventive maintenance, whereby certain equipment failures can be prevented with regular maintenance interventions. Maintenance planning and control also started to evolve to ensure the correct maintenance practices are carried out.

Over the years a number of unfortunate incidents, such as the *Amoco Cadiz*, Bhopal, Chernobyl, and Piper Alpha disasters created widespread public awareness and set higher expectations for many other industries. This led to the emergence of tighter regulations around health and safety, and environmental compliance, and brought authorities to ask for more from the maintenance function. Focus shifted to higher reliability, to greater safety, environmental compliance, and energy conservation. This natural evolution is illustrated in Fig. 1.4 below.

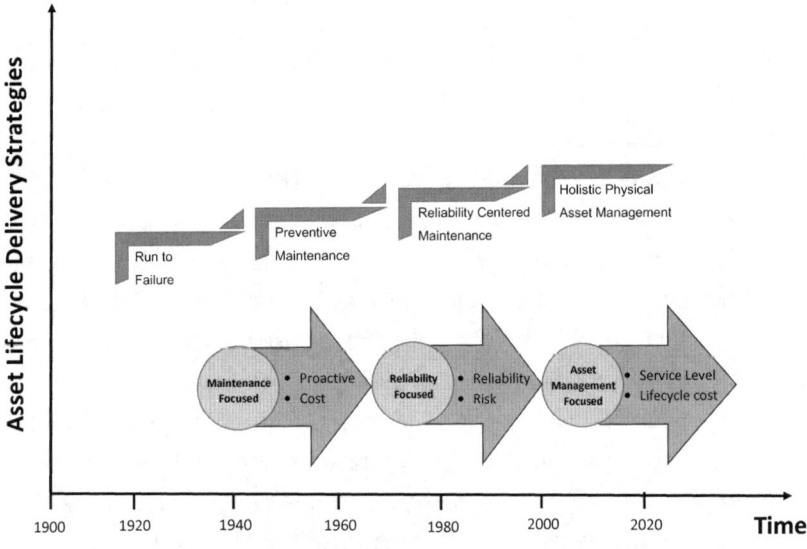

Fig. 1.4 The Evolution of Maintenance & Reliability

Led by the innovation and high standards set in the military, the maintenance function in other industries gradually evolved into reliability, whereby structured approaches were initiated to develop sound maintenance and reliability practices. The aviation industry which was experiencing high risks of failure was the first one to take on the paradigm shift, thanks to Stanley Nowlan and Howard Heap's report, *Reliability-Centered Maintenance*, in 1978. The evolution of and the paradigm shift in maintenance and reliability, toward reliability-centered maintenance (RCM) continued through the 1980s in many other industries such as power plants, utilities, and other manufacturing facilities, thanks to John Moubray from the Aladon Network, who developed and applied the RCM2 methodology. Since then, many RCM methodologies have been developed and rolled out in a wide range of industries.

The Evolution to Physical Asset Management

Over the last several years, we have heard of the need to take a more holistic approach towards physical asset and infrastructure management. Organizations, both in the private and the public sectors, have been doing their best to take care of their ageing physical assets and infrastructure. However, the increasing occurrence of physical asset/infrastructure failure, the demand for more effective management strategies, the continuous lack of funds, and the increasing expectations of service levels have combined to put physical asset/infrastructure management as one of the top priorities for many organizations. When it comes to physical asset care, rather than focusing solely on maintenance and reliability, there is a push to evolve to whole asset lifecycle management. This is required so as to have better asset planning, to minimize risks, and for better decision making to optimize asset lifecycle costs. Every physical asset goes through similar lifecycle phases, from cradle to grave. While different literature describes the lifecycle phases differently, in this book we will stick to the six physical asset lifecycle phases as shown in Fig. 1.5 below.

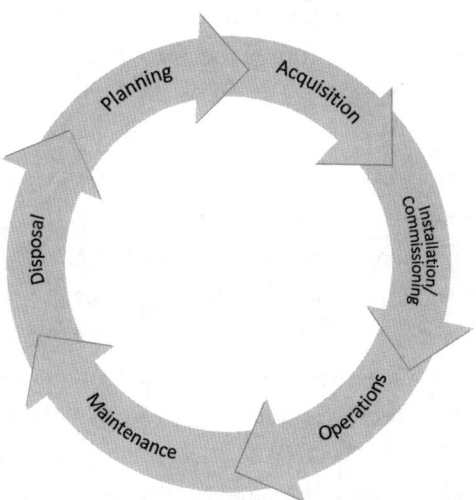

Fig. 1.5 Physical Asset Lifecycle Phases

1. *Planning* covers the activities related to identifying needs, evaluating different alternatives, selecting the right option, allocating funds, and agreeing on timing.
2. *Acquisition* consists of the design and procurement of the identified physical asset.
3. *Installation/commissioning* involves the creation, building, and installation of the physical asset and ensuring it functions as per requirements.
4. *Operations* covers the utilization of the physical asset to realize value.
5. *Maintenance* consists of the tactics required to ensure the physical asset performs all its function safely throughout its economic life.
6. *Disposal* involves removal of the physical asset from operations and its safe disposal.

Each of the physical asset lifecycle phases is very important to the organization, to the people handling them, to society, and to the environment. Failures can happen at any point in time in the lifecycle phases of the physical asset. If these physical assets phases are not properly managed, failures can turn into catastrophic incidents with severe consequences. While the practice of physical asset management has been around many years, what has been missing was a methodical, more strategic, and holistic approach, enabling organizations to maximize value, minimize risk, and deliver on strategic objectives through their physical assets.

Publications such as PAS 55, the International Infrastructure Management Manual (IIMM), and more recently the ISO 55000 suites, have been instrumental in helping to disseminate and share physical asset management best practices and knowledge. Many organizations around the globe—such as The Institute of Asset Management (IAM), The Asset Management Council (AMC), The Institute of Public Works Engineering Australasia (IPWEA), Plant Engineering Maintenance Association of Canada (PEMAC), the Global Forum of Maintenance and Asset Management (GFMAM), the Association of Asset Management Professionals (AMP), Reliabilityweb, and many others—are helping physical asset management practitioners through this evolution. Many organizations, both in the public and the private sectors, have started to adopt the paradigm shift. In a nutshell, the physical asset management paradigm shift is all about aligning the bottom-up approach with the top-down approach for physical assets.

Unlike the transition from maintenance to reliability, which involves at most two functions in an organization (most organizations have maintenance and reliability reporting to the same unit), the transition to physical asset management is a more complex endeavour. This complexity exists because physical asset management:

- is a relatively new field (in its present holistic context) to many organizations;
- is a more strategic approach than maintenance management;
- involves many functions of an organization and requires all of them to align completely and work together.

To create a fully operational physical asset management function in an organization is a daunting task in itself. Without a doubt, maintenance, reliability, and physical asset management have become key functions for any organizations dealing with physical assets. These functions are not new concepts to organizations at all. They have been around for a long time and organizations have been applying these evolving concepts for years, with different level of success and with different effective approaches.

The question here is how well organizations are making the transitions from maintenance to reliability and, to that effect, from reliability to physical asset management.

Is it just a case of rebranding and calling it a different name, wanting to be able to say "Yeah, we're doing it, too!" so we can look like other organizations?

Or is it more of a case of not knowing how to make the transition from one to the other, because we don't know what physical asset management is?

In some cases, it may entail major organizational restructuring, where you would be asking,

"How do I go about implementing physical asset management in my organization?"

Maybe there is genuine intention to make the transition, but the organizational culture is not there, or maybe upper management does not support the transition. All these are good reasons why organizations are not able to make the leap and make the paradigm shift happen.

If we look around in different industries, some organizations are or have been struggling to make the transition from a maintenance mindset to a reliability mindset because of the reasons mentioned above. Transitioning from a reliability mindset to a physical asset management mindset is even more difficult, because now we are not talking about only one or two functional areas in an organization, but multiple functional areas. The core of this book investigates the physical asset management function from an organizational perspective, how it fits, and the challenges organizations are facing to make the leap. The chapters that follow explore why organizations are struggling to make the transition to holistic physical asset management and the challenges they are facing to make it happen.

CHAPTER 2 —
Typical Functional Areas of Organizations

"The whole is greater than the sum of its parts." ~ Aristotle

Organizations and Functions

Organizations are formed to pursue goals that their owner(s), shareholders, communities at large, or board members have identified. In order to maintain and sustain focus, many organizations develop vision and mission statements, along with a corporate strategy. The **vision statement** provides long-term direction and motivation for the organization. The **mission statement** stipulates the organization's business scope and methods of competing. The **corporate strategy** defines the specific businesses or markets in which the organization will compete. Depending on the size and complexity of the organization, their corporate strategy can be broken further down into specific business strategies. So that they can survive and prosper in the current, fiercely competitive marketplace, organizations develop organizational strategic objectives that are aligned with their corporate and business strategies. This enables organizations to consistently produce goods or deliver services that meet their customers' expectations. Organizations can be for-profit or non-profit entities and their goals and products can be similar or different.

Organizations are typically made up of three main components: technical, conceptual, and human. These three components are sometimes referred to as the three Ps: people, process, and plant. All three components are key elements for an organization to prosper and to successfully deliver the fourth P, products as illustrated in Fig. 2.1 below.

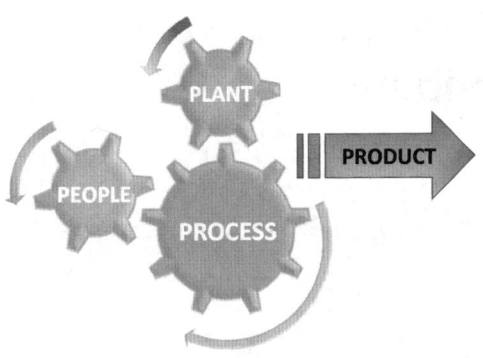

Fig. 2.1 Interaction of Organizations' 4 Ps

The interactions of the three Ps are critical for organizations to deliver the right product in the form of goods or services, and to meet their organizational strategic goals.

People are the individuals and teams of people who perform work, consistent with the organization's values and culture, to achieve strategic goals and objectives. People requirements start with leadership, competencies, knowledge, and accountability at every level of the organization. People are the cornerstone of any organization to accomplish work through shared vision and passion, and are the balancing point between plant and process that enable a business to thrive.

Process includes the organizational structures, systems, and procedures that are used to deliver the product that the organization provides, as well as the policies and rules that support these systems and procedures.

Plant involves the infrastructure, physical assets, and equipment necessary to meet an organization's production or service delivery demands in terms of quantity and quality. Major investment is required in physical assets, from acquisition to operations and maintenance, to maximize their economic life.

Product, the fourth and final P, are the goods or services resulting from the interaction between people, process, and plant. Products are made to satisfy customers' specific requirements.

The interaction of the three components—people, process, and plant—creates the operations function that performs all the activities related to producing goods and delivering services. Operations is one of the three main functions of a typical organization. The other two are finance and marketing. These three functions interact with each other, and together perform different but related activities to help the organization achieve its strategic goals, meet its customers' demands, and remain competitive in the marketplace. The ***operations function*** is the core of any organization. Its interface with finance is critical to ensure funding needs are met, and the required physical assets and materials are available when needed. Similarly, interface with marketing is necessary to ensure operations can meet the demands and requirements of the market, when needed.

Besides these three main functions, and depending on the complexity and/or size of organizations, other functional areas may be required to support the three main functions, to manage more effectively and efficiently the three Ps, and to harmonize the relationships and interactions between all functions.

The most common supporting functional areas are human resources, information technology (IT), purchasing, accounting, maintenance, project management, and engineering. There may be other functional areas, depending on the nature of the business and the organization's structure.

The functional areas, in general, perform the activities necessary for the effective and sound management of the organization's transactions. These functional areas bring technical, conceptual, and human components together and make them gel to create value from the organization's assets as shown in Fig. 2.2 below.

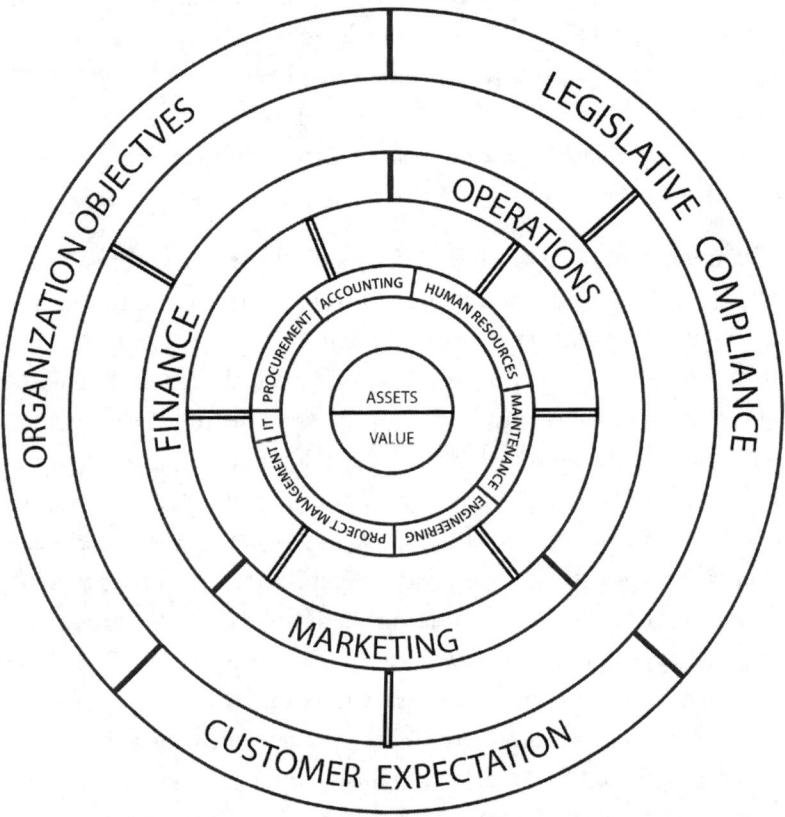

Fig. 2.2 Traditional Functions of a Typical Organization

Make no mistake: each function is equally important and brings great value to the organization. However, the functions must interact and work together to achieve the organization's common goals and objectives. For example, operations and purchasing must work closely together to order raw materials at the right price, the right quality, and at the right time. Similarly, maintenance must work closely with finance to ensure the funds are available for their next major plant shutdown.

Let us now take a brief tour of the traditional functions of a typical organization to better understand these functions and their mandates.

Operations is the core function of an organization. It is the process of transforming inputs such as raw materials, labour, machines, etc., into outputs such as goods and services as specified by customers. The essence of the operations function is to add value during the transformation process.

Finance's mandate is to ensure the necessary funding is available at the right time to secure resources at favourable prices. To make that happen, careful planning and accurate budgeting are required to avoid cash flow problems. Finance is also responsible to carry out economic analysis on investment proposals and look for proper sources of funding.

Marketing is responsible to identify and assess customers' needs and wants, and to translate those into goods or services, working with operations and the design department. Marketing's mandate is to scan the market and identify opportunities for accurate sales forecasting and operations scheduling.

Accounting collects and supplies information to management on costs of labour, materials, and overhead, and provides reports on items such as inventories, downtime, and scrap. Accounting deals with information needed for financial statements and carries out internal audits to ensure adherence to accounting policies and regulations.

Procurement has responsibility for the procurement of materials, supplies, equipment, and services. They interact closely with operations to guarantee correct quantities and timing of purchases. They also ensure organizations are getting value for money from their vendors in terms of cost, quality, delivery-time reliability, service, and flexibility.

Maintenance is responsible for general upkeep and repair of equipment, buildings, and other physical assets. Their main purpose is to provide safe and reliable equipment and infrastructure to support operations by applying sound maintenance strategies on a regular basis.

Project management is responsible for delivering capital projects, as and when required by the organization. Its main purpose is to ensure capital projects are managed from start to finish, and that they meet all the specified requirements and performance standards, through the application of sound project management methodology.

Engineering is responsible to provide technical/engineering skills, carry out engineering studies, and develop design standards and specifications. Its primary purpose is to ensure the organization is in compliance with applicable engineering regulations and standards.

Information technology is responsible to store, retrieve, transmit, and manipulate data using computers and telecommunication equipment. Organizations require historical and up-to-date data to be accessible all the time, as well as being able to leverage technologies to retrieve, transmit, and manipulate data.

Human resources is responsible for hiring, developing, evaluating, rewarding, and retaining staff, based on skills and qualifications, to perform specific tasks in the organization. Its mandate is to develop strategies to ensure organizations are properly staffed, with the right people with the right skills and values to fit the organization's culture.

So far we have seen the different functions/functional areas that support an organization in its quest to meet its strategic objectives. We have reviewed the mandates of each of those functional areas and have also highlighted the fact that organizations may have other functional areas, but the ones mentioned above are the most common ones.

"So what is the common thing that all those functional areas are trying to control and manage in an organization?" you ask.

The answer is simple: the organization's **assets**.

Let us now take a closer look at the assets of organizations, which are the center of attention of the functional areas.

Assets in Organizations

Asset is a term that is broadly used in different industries and organizations, and assets come in different shapes and forms. As defined in the ISO 55000 standard: "An asset is something that has potential or actual value to an organization." All organizations deal with different types of assets that are necessary to achieve organizational strategic objectives and hence bring value. Assets in organizations can be classified into five distinct categories:
1. *Tangible assets* are those assets that we can touch, see, and feel. These are buildings, infrastructure, equipment, inventory and properties owned by the organization. They are used primarily to produce goods and deliver services to meet customers' requirements.
2. *Intangible assets* cannot be seen, felt, or touched physically. Some examples of intangible assets are goodwill, franchise agreements, patents, copyrights, brands, trademarks, etc.
3. *Human assets* are the organizations' owners, managers, employees, contractors, and suppliers. They are the workforce of the organization.
4. *Financial assets* are the bonds, securities, stocks, or bank deposits. The organization uses its financial assets to fund operations, maintenance, materials, and infrastructure investment.
5. *Information assets* are the body of knowledge of an organization that is organized and managed as a single entity. They are usually in the form of good quality information and data that is used to make business decisions and develop future strategies.
6. *The meaning of value* will vary depending on the industry, the organization itself, and its stakeholders. Value generated from assets can be tangible or intangible, financial or non-financial.

The relation and interaction between the different types of assets are key factors for the creation of value. Hence, effective control and governance of those assets becomes very critical for any organization to reap the benefits from its people, process, and plant and to fulfill the organization's objectives. The various functions in organizations are tasked to do exactly that; i.e., to manage and control the assets, with the three Ps of the organization, to deliver maximum value. For each functional area in an organization, value will be measured or perceived differently.

Organizations hold varied importance to their assets, depending on their criticality with respect to the industry, the nature of the organization, the operating context, the regulatory requirements, or stakeholders' expectations. For example a mining company will take care of its trucks to ensure they are available to transport products from the mining site to the processing facility, a banking institution will manage its financial assets for short-term profit-taking, a manufacturing plant will manage it's workforce to ensure they are motivated and fully engaged, a software company will manage its software licenses to prevent piracy or the Jurassic World theme park will manage its assets (for example, the *Indominus rex*) to attract and entertain visitors and tourists on Isla Nublar.

"How interesting!" you say with excitement. "Can we talk a little bit more about the other assets in Jurassic World, such as the Tyrannosaurus and Mosasaurus?"

I can understand your excitement, but unfortunately the main focus of this book is on tangible physical assets.

Physical Assets

Physical assets are acquired for continued and long-term use in operating a business. They include property, facilities, land, buildings, infrastructure, computer servers/accessories, equipment, and machinery that public and private organizations own and manage.

Physical assets can be classified as discrete physical assets or linear physical assets. **Discreet physical assets** have clear and unique boundaries and can be a bridge, a compressor, a conveyor, a mixer, or a vehicle. ***Linear physical assets*** are continuous with one or more undefined boundaries, such as water pipe systems, railway lines, sewer lines, and electrical networks.

In much literature, and in different industries, you will encounter the terms equipment, facilities, and infrastructure being used to refer to physical assets.

- ***Equipment*** refers to industrial equipment, information technology devices and machinery such as compressors, conveyors, servers, vehicles, and mixers. These are widely used both in the private and public sectors and in many different industries.
- ***Facilities*** covers plants, condominiums, office buildings, schools, universities, hotels, hospitals, libraries, etc. Both the public and private sectors own and operate manufacturing/processing facilities and different building facilities to provide other services.
- ***Infrastructure*** refers to the fundamental facilities and systems serving a country, city, or area. The term infrastructure is used widely in municipalities and other levels of government to refer to their physical assets. It typically refers to technical structures in:
  - Transportation, such as roads, airports, ports, bridges, tunnels, and rails
  - Utility and energy, such as water, power generation, electricity, and gas networks
  - Communications, such as transmission towers
  - Social community, such as education, recreation, and healthcare facilities

As per the definitions above, infrastructure can be either discreet or linear.

"Why are we elaborating on the terms used in the industries?" you ponder.

Well, the reason is mainly to highlight the different approaches and challenges of the public and private sectors with respect to physical assets. Understandably the public sector deals predominantly with infrastructure while the private sector deals predominantly with equipment, and both deal equally with facilities.

So let us take a look at the differences and similarities between equipment, and infrastructure. The major difference is that infrastructure has a relatively longer operational life than equipment, and hence the control and governance approach is somewhat different. This brings us to introduce the concepts of equipment-centric and infrastructure-centric organizations to help us better understand control and governance approaches.

Equipment-Centric Organizations

Equipment-centric organizations are those organizations that have high numbers of complex and critical equipment, such as pumps, conveyors, computers, mixers, boilers, and electrical systems. These types of equipment have relatively shorter lives, suffer frequent wear and tear, and hence require intensive care and attention. Most equipment-centric organizations are heavily involved in activities to manage wear and tear, and to extend the economic lives of their physical assets. These are the tactical activities that are performed on a routine basis to ensure high physical asset performance and reliability. During that time, many organizations get caught in this tactical battle and tend to neglect the strategic activities that need to happen. The results are reactive situations, poor decision making, poor physical asset planning, high lifecycle costs, and recurring physical asset failures. These are some of the most common signs of equipment-centric organizations that are embroiled in what is called the spiral of death.

"Interesting," you observe. "Now I understand why we find so many organizations struggling to collect relevant data on physical assets, to build physical asset history, and to analyze lifecycle costs. They are so much focused on the short-term tactical activities that they lose sight of the need for physical asset data analysis for better decision-making."

Correct. That is why these organizations are very strong in planning tactical activities such as proactive maintenance, but somewhat weak in planning strategic activities such as physical asset upgrade or replacement.

"But why is that so?" you question.

It is simple. Equipment-centric organizations do not proactively put forward capital investment business cases for strategic activities for three reasons. First, they lack the required data to build strong enough cases. Second, they have few resources dedicated to work on the business cases in a timely manner. And, third, organizational policies and strategies for decision making do not support the business cases

Infrastructure-Centric Organizations

Infrastructure-centric organizations, on the other hand, have critical physical assets of different complexities, which have lower wear and tear rates and much longer lives. Hence, care and attention is less intensive and less frequent, even though the relatively less frequent interventions are very crucial and demanding. Failure of physical assets in infrastructure-centric organizations can cause catastrophic fatalities or major disruption to essential services. They are built to last longer life and designed to incorporate room for future expansion and growth. Examples of such physical assets are roads, bridges, electrical grid systems, and buildings.

"I think I am beginning to understand," you say. "Am I correct in saying that those organizations have more time and dedicated resources to focus on collecting and analyzing data for better future planning?"

You've got it! Infrastructure-centric organizations dealing with those physical assets do a great job of assessing conditions on a regular basis and of planning for refurbishment or replacement, since the tactical activities are less frequent. This is critical because those physical assets (infrastructure) are typically much more expensive than the equipment-type physical assets and hence proper planning for future capital investment is crucial.

From now on, to simplify things we will include everything—infrastructure, equipment, and facilities—under one umbrella as physical assets.

The Physical Asset Management Function

Equipment-centric and infrastructure-centric organizations need to take a more holistic approach when it comes to managing their physical assets. Physical asset management is another functional area that many organizations are adopting to ensure their physical assets are taken care of, both tactically and strategically. Physical asset management is responsible for minimizing the lifecycle costs of owning, operating, and maintaining physical assets, at an acceptable level of risk, while continuously delivering established levels of service. Its main purpose is to ensure all the tactical and strategic activities required throughout the whole lifecycle of physical assets, from cradle to grave, are taken care of to ensure organizational strategic objectives are achieved. It is, in fact, a balancing act for optimal decision making to find a compromise among cost, risk, and performance.

"Excellent!" you exclaim. "But that's not easy to achieve. As per my experience, many organizations have been working hard to achieve this for many years, with various degree of success."

In a sense, that is very true. Physical asset management has been practiced for years, but it is only in the last five to eight years that we have seen an emerging demand for a more holistic and structured approach. For many organizations, this is not an easy feat in itself. The physical asset management function has been facing a tough ride to find its berth as a recognized functional area in organizations. As mentioned earlier in this chapter, evolution toward holistic physical asset management has to happen to tackle the issues of ageing physical assets while providing the required service level at lowest risk. But along the way we have to recognize that there will be many organizational challenges to overcome. Challenges can range from fitting the physical asset management function in organizations that already have several other well established functional areas, to having to change the entire culture of an organization to embrace a new philosophy with new ways of doing business.

In subsequent chapters, we will explore the challenges faced by each of these functional areas with respect to the physical asset management function and vice versa.

CHAPTER 3 —
Physical Asset Management's Fit in Organizations

*"The achievements of an organization
are the results of the combined effort of each individual."*
~ Vince Lombardi

In Chapter 2, we saw that organizations have different functional areas in place to ensure physical assets are taken care of over their entire lifecycles, to bring value to the organization. We have also seen that organizations have different types of physical assets to manage, with targets to ensure required service levels are met, to optimize costs, and to minimize risks. However, what we see in the real world is that all the different functional areas of organizations have different and sometimes conflicting mandates. As a result, the functional areas are not able to fully align their efforts in managing physical assets for the benefit of the organization. The fact that holistic physical asset management is a relatively new function in many organizations (even though organizations have been practicing conventional asset management for decades) it is currently facing a significant challenge to fit into the existing organizational hierarchy, and coexist with the other functional areas.

Fig. 3.1 "Can you move over, please"

"But why would physical asset management face such challenge to fit in?" you ask. "Don't organizations realize the importance and benefits of holistic physical asset management?"

Yes, indeed. Many organizations in the public and private sectors understand and realize the importance and benefits of physical asset management. While many are investing heavily to start implementing it, the reality is that many organizations are simply not ready for the undertaking. They are either struggling to accept the philosophy or to reengineer themselves, for the following reasons:

- Some organizations believe they are already practicing physical asset management via existing functions.
- Existing functional areas operate in silos, with different and conflicting goals and objectives.
- Some organizations believe that physical asset management is already covered by their maintenance management function.
- Other organizations think that they do not need the physical asset management function at all.

These organizational beliefs or culture make it difficult for them, as a whole, to embrace the shift to holistic physical asset management.

Before we analyze the challenges that physical asset management is facing, let us first take a look at organizational anatomy and design to better understand how organizations operate as a whole.

Organizational Design

Organizations are tasked to coordinate the work of different functions via horizontal and vertical integration to meet organizational strategic goals and achieve operational effectiveness. This coordination task takes the form of a formal, guided process for integrating people, information, and technology within the organization, as well as including the division and coordination of work activities. There are typically four key areas in which organizational design takes place to achieve strategic goals and operational excellence:
- Work specialization
- Allocation of authority
- Departmentalization
- Span of control

The degree of focus in each of these key areas determines the organizational design approach, ranging from mechanistic to organic.

Mechanistic organizational design is a centralized structure where there is a high division of labour and low delegation of authority, with boss-centered decision making. This design promotes tight rules, emphasizes specialization, and limits employees' discretion. It restrains collaboration across the organization.

Organic organizational design is a decentralized structure where knowledge is shared through cross-training for specialization, where decentralized control and decision-making prevail, where empowerment and participation are encouraged. This design promotes collaboration, communication, and widened span of control.

Creating a physical asset management function in an organization will help to shift some of the mechanistic approach vis-à-vis physical assets to a more organic approach. The shift will be especially important and noticeable in the four key areas, whereby collaboration and communication will be enhanced across different functions. Empowerment and participation will be encouraged at all levels of the organization for optimum decision making.

If we take a look at organizations today, we see that this is not the case. Interactions between different functional areas vary tremendously depending on their sizes, their complexities, and how the organization is structured. Organizational structure reflects a firm's competitive strategy, where its senior leaders review and adjust division of labour, delegation of authority, departmentalization, and span of control to align with the organizational strategy.

In a centralized structure, the decision-making power is concentrated at senior management level, and there is tight control over the functional areas. In a decentralized structure, the decision-making power is distributed, and the functional areas have varying degrees of autonomy, which allow them to be more flexible and better interact with each other.

Organizational design and organizational structure are key factors that dictate how different functional areas collaborate and interact with each other, how decisions are made, and how control and authority are defined. Many organizations are making considerable effort to move from a mechanistic to a more organic approach, from a centralized to a decentralized structure, so as to encourage healthier collaboration and communication between functional areas and to streamline efforts to achieve organizational strategic goals.

Leadership and Organizational Culture

A lot also depends on the leadership and the culture of the organization. **Leadership** is the influence process in which one person guides, directs, and channels the actions of another.

"We're talking about the CEO or the senior manager of the organization," you point out. "Is that correct? But what about the rest of the organization?"

Excellent question! Leadership in an organization is not confined to the executive or top manager only. While we recognize that leadership at the helm of an organization is critical to crafting a new vision for the organization, to steer it in a new direction, and to convince employees to go along with it, leadership also needs more horsepower to drive the change required to become successful. When an organization transitions from a centralized to a decentralized structure, it empowers and engages everyone at all levels to participate. In the process, it needs to grow and develop the leadership talent of every single employee throughout the organization to help drive the influence process, to accept and drive change, and to apply new methods to old problems. This requires lots of efforts on behalf of the organization to ensure the right leadership is in place. In many instances, organizations are stuck in the mechanistic approach and are trapped in a leadership conundrum while they continue to struggle to streamline people's efforts to align with organizational objectives. The organization's leadership must ensure that the responsibilities and authorities for interpersonal, informational, and decisional roles are assigned and communicated within the organization to promote cross-functional collaboration and maintain mutual trust and high performance.

"But what about the culture of an organization?" you ask.

Another very good question, and this time the timing could not be better!

If you recall, we said that leadership is not confined to executives or senior managers, but applies to everyone at every level of the organization. We also said that organizations need more horsepower (more leaders) to influence and drive change in the direction set by the organization's vision and mission statements. People in different organizations, different industries, and countries work differently, think differently, communicate differently, behave differently, and make decisions differently. These differences shape the different cultures within organizations.

Organizational culture is a difficult concept to define and you may get different answers from different people. In a nutshell organizational culture is a set of shared mental assumptions that guide interpretation and action in organizations by defining appropriate behavior for various situations. In organizations, organizational culture is the collection of business practices, processes, and interactions that make up the work environment. Because of variation in work environment and geographical location, organizations develop business units and functional areas with distinctive cultures and subcultures; i.e., with their own beliefs, assumptions, and styles. While the development of cultures and subcultures in organizations and functional areas is an unavoidable phenomenon, it is very important that these cultures/subcultures synchronize well with the organization's strategy. Over time, the organization's strategy and objectives may change, and subsequently, it is essential for organizations to reassess the fit of the current culture/subculture and to undo aspects of the organizational culture that reinforce old practices. All adjustments of organizational culture to meet the expectations of new strategic objectives must be done within the realm of the organization's values.

"But organizational values are the same as organizational culture," you reply.

Actually, no. Organizational culture and organizational values are completely different things. Organizational values are the set in stone core principles that the organization is willing to abide to; they are the rules of the game. Strong values will guide the decision-making process and give a sense of what's important and what's right for the organization. An organization needs to have strong values to unite people across the organization, no matter in what functional areas, type of organizations, or industries they work in and wherever they are in the world. All those people are in pursuit of the same organizational objectives, and will build on the foundation of the company's values to achieve those objectives. Let us say that organizational values are the roots of a tree and, the stronger they are, the better it is for the organization. Culture, on the other hand, is the actual branches and leaves of the tree, which have to be nurtured and taken care of to maintain a motivating and pleasant work environment. Branches and leaves change over time due to seasonal changes, natural disaster or human actions. Culture is the actual realization of the organization's values driven by the needs of the organization over time.

So far we have seen that leadership and culture/subculture are built on a solid foundation of organizational values. They are key drivers to promote cross-functional collaboration and high performance in an organization. The physical asset management function—which relies heavily on interactions between functional areas for collaboration and decision-making—is in a sense trapped in the conundrum of leadership and culture/subculture. Because of that, physical asset management is finding it hard to "fit in" within the existing organizational structure, which then affects its ability to deliver and to make a difference.

"But why is that so?" you quickly ask. "How come other functional areas can fit in easily and why not physical asset management?"

This is the multi-million-dollar question and, basically, what this book is all about.

Let us take collaboration as an example of an organization value. Good collaboration in an organization means the right people in different functional areas working together to deliver a common product or to make a decision. Now, if that collaboration is poor or not well managed, it will mean a constant struggle to get buy in or to work together to come to the common decision. Culture needs to change over time in order to promote and enhance collaboration in an organization. In small organizations collaboration is more easily achievable. On the other hand poor collaboration cannot be felt as much in larger organizations, where there are more subunits and functional areas with different mandates and areas of focus, thus creating more subcultures and differences. For collaboration to be effective, all these functional areas must work together with a common focus and set aside cultures/subcultures differences.

Areas of Focus of Functional Areas

Because of variation in work environments and geographical locations, organizations develop business units and subunits to support the functional areas to balance between competitive strategy and organizational design. To manage those business units/subunits (we will use functional areas as generic term) many organizations adopt a results-oriented control system for example similar to management by objective (MBO). MBO achieves results-oriented control because it requires managers from each functional area to set specific goals and desired targets in the strategic plan. Within each functional area, managers monitor and reward their staff for achieving those defined targets, while allowing staff to select their own methods to work towards their goals. Besides focusing attention on output control, MBO develops an organization's culture of high standards, staff involvement, and participative decision making. Very often, this culture can become result-obsessed and experience internal competition, which results in poor collaboration and coordination among staff and between the various functional areas. Everybody has their own marching orders, and has their own missions either to look good or to make their manager/functional area look good. Once organizations reach such a state or culture, it becomes a significant challenge to realign them and their functional areas to work together towards common organizational goals.

"Are you not referring to organizational silos, by any chance?" you ask inquisitively.

Right on! This is exactly what I was coming to—the famous "bureaucratic silos" that we know exist in many organizations, the silos created by those functional areas that are left to operate in their own worlds. Silos in organizations are created by the organization itself and by its own leaders. Initially, the different functional areas were designed and structured to become an integral part of the organization, with a common goal or common goals (strategic objectives). Instead, the functional areas end up operating as distinct business units, with different and very often conflicting goals and objectives. Organizations who fail to bring those functional areas together and realign them end up being non-effective organizations with lots of internal issues. We fully understand that each functional area is different; each is as important as the other, each requiring different expertise and knowledge base, and each having different challenges. Functional areas tend to be more focused in resolving their own challenges rather than working towards resolving the organization's challenges and achieving organizational objectives.

Fig. 3.2 illustrates a schema of the triangles of focus (ToF) of functional areas to illustrate how different the areas of focus of an organization's functions can be.

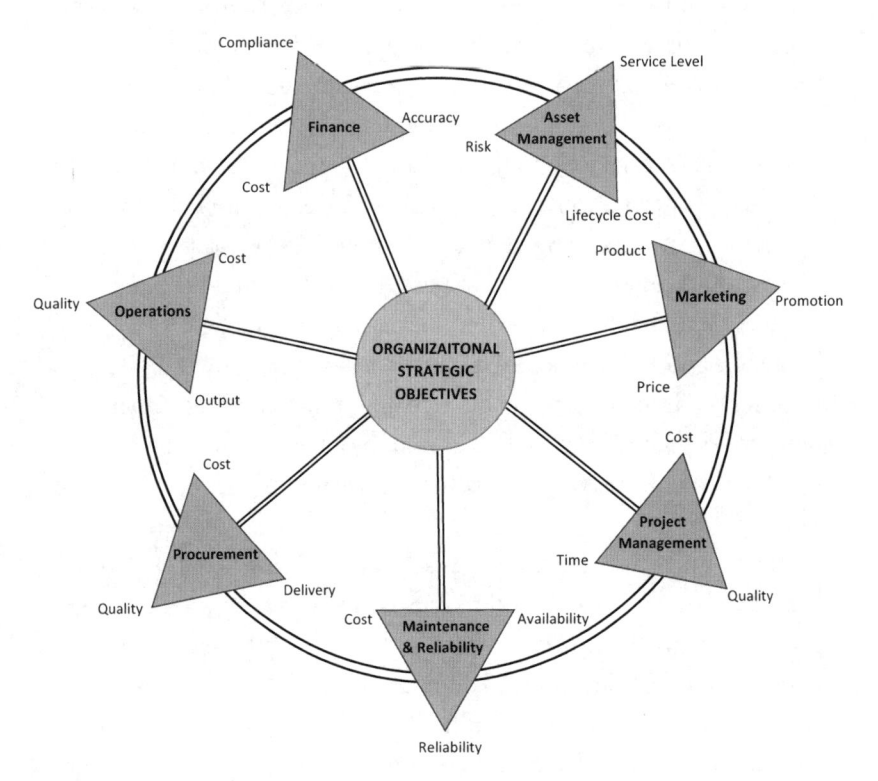

Fig. 3.2 Triangles of Focus of Functional Areas

The ToF is to identify only the top three areas of focus for each functional area in an organization, even though some may have different ones or maybe more than three areas of focus. Many of you, if working in those functional areas, will surely recognize those areas of focus, either on your performance planner or in the form of performance metrics that you measure and track on your dashboard. Each functional area is managed with specific objectives and goals, which are represented by the ToF.

"I notice that some areas of focus are completely different, but many others are similar," you reply.

Good observation. Understandably, you will find many areas of focus from different functional areas that are similar to each other. However, that does not necessarily mean that the functional areas are totally aligned. Let us take "cost" for example. Cost is an area of focus for both the procurement and the maintenance functional areas. For the procurement function, this means procuring physical asset parts, components, and materials at lower costs. In many organizations, this is achieved via lowest bid procurement systems. This can have an impact on the quality and reliability of the parts, components, and materials, which can affect the performance of the maintenance function. Now the maintenance function is also focusing on cost, but will achieve low cost via different tactics, such as better planning, scheduling, and maintenance task optimization, to minimize physical asset failures and improve reliability. So, on one side, the maintenance management function will require better quality and more reliable physical asset parts and components, which may cost more, while on the other side the procurement function will be selecting similar items for the lowest cost, sometimes at the expense of quality and reliability.

"I see what you mean," you declare. "I have seen this type of situation before. So how do the functional areas resolve this?"

Well, they have to work closely together to find the right balance. Both functional areas need to look into the details to determine and agree on where and in which circumstances cost, quality, or reliability is more important for the organization – not necessarily for their specific functional areas.

However, the situation is totally different when you have functional areas with completely different areas of focus.

"But hold on a second," you interject. "Areas of focus must be different because they are from different functions with different mandates, different knowledge bases, and different goals."

Of course, I totally agree with you. Different functional areas will obviously have different areas of focus. The key here is how each functional area manages and achieves those areas of focus. Are they achieved without consideration for their impact on the other functional areas or on the organization as a whole? To illustrate the situation, let us take the example of the project management function with "time" as an area of focus. We all know how meeting deadlines is critical for projects. However, we also know that unpredictable stuff sometimes happens and projects get delayed or extended. However, what may be acceptable for projects may not necessarily be convenient for the operations function. For example, in the operations functional area, one key area of focus is "output." If the project is delayed and not delivered on time, it can have an impact on the production schedule and hence on the production output.

Even though we recognize that different functional areas will have different areas of focus, they still need to work together to ensure those areas of focus are not achieved to the detriment of other functional areas, and the organization as a whole.

Having said that, just imagine for a moment how it would look if all functional areas of an organization had all different areas of focus! How about all the functional areas having all the same areas of focus?

The point I am trying to make here is that, in any organization, you will find a mix of different, similar, aligned, and non-aligned areas of focus. This various mix in areas of focus will create gaps and overlaps (G&O) between the functional areas. These G&O are one of the reasons we have functional silos in organizations. Organizations need to find ways to bridge those gaps and streamline those overlaps as best as they can. Functional silos will exist, but they must be harmonized functional silos that are perfectly synchronized with each other to achieve the same organizational goals. In Chapter 4, we will talk more about harmonized functional silos.

Line of Sight

For now, let us see how organizations deal with the various areas of focus, whether they are different or similar. This brings us to the term *line of sight*. Organizations need to find and clarify the line of sight for each functional area.

"What does that exactly mean?" you ask.

This means that organizations need to understand how the goals and objectives of each functional area help to achieve the organization's strategic goals and objectives.

"Shouldn't it be the other way round?" you question. "I believe each functional area needs to understand how their goals and objectives are contributing to achieving the organization's goals and objectives."

You are also correct. The process should be both ways—a top-down enlightenment as well as a bottom-up adjustment. It is crucial for functional areas to refocus on their respective goals and objectives to ensure these are aligned with organizational strategic objectives. Very importantly, let us not forget about those physical assets that contribute to achieve these organizational goals and objectives. Are we really doing the right things to them and taking the right decisions to make sure they continue to bring value to the organization? The line of sight traces a top-down and bottom-up approach of linking organizational high-level goals and objectives to each functional area's goals and objectives, through to physical assets' performance. Talking about performance—each functional area must define performance measures for their goals and objectives, which typically revolve around the physical asset's performance

46 Physical Asset Management: An Organizational Challenge

and service level, which then eventually roll-up to the high-level performance measures for the organization's strategic objectives. Defining and clarifying line of sight is of the upmost importance for organizations to make sure all its different functions work together in harmony and to make full use of their physical assets to achieve their strategic goals in the most effective possible manner.

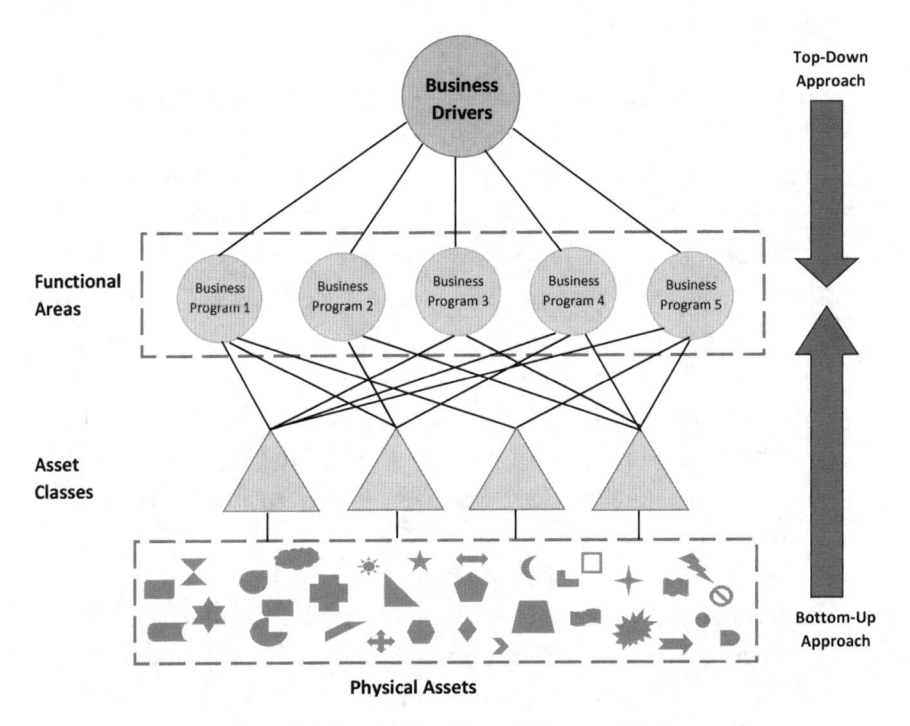

Fig. 3.3 Line of Sight with Top-Down & Bottom-Up Approach

As shown in Fig. 3.3, organizations have different types of physical assets that are utilized to drive and support business programs. The physical assets are regrouped into different asset classes. All functional areas of the organization dealing with those physical assets, directly or indirectly, during their lifecycles must co-exist and work together to support the business programs. Each functional area is important for the organization and has an important role to accomplish along that line of sight. There should be horizontal and vertical involvement, and interaction between the functional areas to ensure physical assets are fully utilized to help the business programs bring value to the organization. The physical asset management function provides some of that synergy. Physical asset management bridges the gaps and addresses the overlaps between the functional areas, with focus on achieving organizational objectives via holistic management of the physical assets.

"But this is not as easy as you are making it sound!" you exclaim. "The gaps and overlaps span across the whole organization, and it's a huge undertaking to address all of them."

Well, you are absolutely right. This is easier said than done. As a case in point, we can see many organizations still struggling to keep their physical assets in a state of good repair to continue to deliver goods & services at optimum cost and minimum risk, even though physical asset management has been practiced in some shape or form for many years. Yes, ageing infrastructure, deteriorating physical asset conditions, and limited funds are indeed major contributing factors in this struggle. But we should not discount the fact that many organizations and their stakeholders have not been up to the task and have been slow to react as an entity to alleviate the situation.

However, the increasing occurrence of physical asset failure, the demand for more effective management strategies in today's tough economy, and the focus on regulatory compliance from organizations have combined to put physical asset management on top of many boardroom agenda.

Who does not remember the Piper Alpha explosion, the Bhopal disaster, the Deepwater Horizon catastrophe, and the numerous sewer and bridge collapses around the globe and their subsequent aftermath?

"Yes, I am aware of those unfortunate accidents," you reply. "Even nowadays similar incidents are still taking place in different parts of the world."

Sadly enough, the risk associated with physical asset failures has been a main talking point over the last several years and has become one of the key preoccupations of many organizations. There are growing concerns about regulatory compliance from governmental and industry institutions. There are growing expectations from consumers about quality and service delivery, and there is more emphasis on cost optimization. All this has contributed into the introduction of a holistic physical asset management philosophy to ensure that organizations have concerted and structured efforts in place to make the right decisions and take more proactive care of their physical assets.

Support

Well, here is the good news. There are many helping hands that have come to the rescue of government, corporate, and non-profit organizations to help them get out of this physical asset vicious circle. Over the years, different standards and publications such as IIMM, PAS 55, the ISO 55000 suite, the IAM Conceptual Model, the AMC Delivery Model, and the GFMAM landscape have been introduced to help organizations fit the holistic physical asset management philosophy and function into their organizational structure and strategy. Obviously there are way more publications out there which are equally beneficial and helpful. We are very grateful to these organizations recognizing the need and having taken the initiative to develop these documents, which facilitate the task of senior managers and asset management practitioners to fit the physical asset management function into their respective organizations. What follows is a brief overview of those standards and publications mentioned above:

The **IIMM** was first published in 2000 and is the result of a huge international collaborative effort from experts in the physical asset management field from New Zealand, Australia, the United Kingdom, the United States, and South Africa. The manual provides guidelines for organizations to manage all areas of infrastructure risk, as well as guidelines to develop and operate sustainable physical asset networks and deliver the required services to our communities at the lowest lifecycle cost. It also contains a brief overview of the theory, how to get started, and what would be expected in a first physical asset management plan. All of this is supported by numerous case studies tailored to the size and complexity of different organizations.

The **PAS 55** asset management standard was first published in 2004, with a second publication in 2008. The standard provides specifications and guidelines for the optimized management of physical assets. Its requirements and structure are arranged in seven key elements within the Deming's Plan-Do-Check-Act (PDCA) cycle:

1. General Requirements
2. Asset Management Policy
   2.1 Asset Management Strategy, Objectives and Plans
   2.2 Asset Management Enablers and Controls
   2.3 Implementation of Asset Management Plan(s)
   2.4 Performance Assessment and Improvement
   2.5 Management Review

According to the PAS 55 asset management standard, *asset management* can be defined as:

> Systematic and coordinated activities and practices through which an organization optimally manages its physical assets and their associated performance, risks and expenditures over their lifecycles for the purpose of achieving its organizational strategic plan.

The above is very much in line with what we have been discussing so far. It is applicable to any organization where physical assets play a key and critical role in achieving business goals.

Another key publication is the **ISO 55000** standard, which was released in January 2014, after almost three years of work by thirty participating countries. The International Standard specifies the requirements for the establishment, implementation, maintenance, and improvement of a physical asset management system. The ISO 55000 standard consists of three suites of documents:
1. ISO 55000: Asset Management — Overview, principles, and terminology, which provide an overview of physical asset management and the standard terminologies and definitions used.
2. ISO 55001: Asset Management — Management Systems — Requirements that specify requirements for a physical asset management system within the context of the organization.
3. ISO 55002: Asset Management — Management Systems — Guidelines for the application of ISO 55001, which explain and clarify the requirements specified in ISO 55001 and provides examples to support implementation.

The other great resource that is of tremendous help to organizations to better understand how the physical asset management function fits in their structure and what the discipline of physical asset management is all about is the **GFMAM Asset Management Landscape**. The Asset Management Landscape is a GFMAM initiative, first published in 2011, to facilitate the exchange and alignment of maintenance and asset management knowledge and practices. The core of the Asset Management Landscape is the thirty-nine subjects to describe the complete scope of asset management and to develop a common approach to asset management. The thirty-nine asset management subjects are documented under six subject groups:
1. Asset Management Strategy and Planning
2. Asset Management Decision-Making
3. Lifecycle Delivery Activities
4. Asset Knowledge Enablers
5. Organization and People Enablers
6. Risk and Review

Going back to how physical asset management fits in an organization, it is worth mentioning two very important conceptual models. These models help organizations create an identity for the physical asset management function and position it in an organizational framework, to show how it contributes in achieving organizational strategic objectives.

The **IAM Conceptual Model** depicts how organizations should be lining up their key physical asset management activities to align with the organization's goals. It is designed to describe the overall scope of physical asset management and high-level groups of activity that are included within this discipline. It helps to demonstrate how those high-level groups of activity are integrated into the big puzzle to direct, coordinate, and control the key physical asset management activities.

The **AMC Delivery Model** is another useful model which schematically presents processes within a number of disciplines that may be used in-part, or in their entirety, to deliver successful physical asset management. It analyzes each step of a physical asset lifecycle from planning, acquiring, and operating/maintaining to disposal, and identifies and maps the key physical asset management activities to deliver effective asset management.

Forward-looking and proactive organizations embarking on the journey of physical asset management must leverage these resources and other publications (including this textbook, of course) to:
- Understand how physical asset management should fit in their organizations.
- Identify and understand the different physical asset management activities that they need to focus on.
- Have a clear idea how all physical asset management activities roll up to contribute to the organizational objectives.
- Develop a physical asset management framework to meet their needs in their operating context.

"Well, that's easier than I thought it would be," you say with relief.

Not so fast. The truth is that, in many major organizations, the journey to holistic physical asset management is not that rosy. Along the path, there are many challenges to deal with. The two main challenges are:
- To fit the physical asset management in an organization as a functional area.
- To drive the paradigm shift across the organization.

In the chapter that follows, we will explore the challenges encountered by physical asset management.

CHAPTER 4 —
Challenges of Physical Asset Management

"What got you here won't get you there."
~ Marshall Goldsmith

What got you here won't get you there! Yes, this adage from Marshall Goldsmith's book of the same name applies not only to people, but also to organizations. So far we have discussed how the physical asset management function could fit in an organization and find its place among the many other existing functional areas, with their different areas of focus. We have also looked at the different conceptual models and standards of physical asset management that help guide organizations in introducing and integrating the holistic asset management philosophy. No matter how the physical asset management function is approached by organizations, there are challenges to be addressed along the way to make it fit into the organizational structure and fulfill its mandate. In the Asset Management ISO 55000 Standards this is referred to as "Relationships between key terms". The key terms are the Asset Portfolio, the Asset Management System, Asset Management and Managing the Organization. Here we are referring to the challenges between the two outermost layers of this relationship, namely **Managing the Organization**, which is the leadership, culture, and management of the whole business, and **Asset Management**, which is the actual coordinated activity of an organization to realize value from its physical assets.

"What are those challenges?" you ask. "Why do they exist? Are the challenges an organizational issue, or are they more to do with the organizational culture?"

Wait a second. That's too many questions at once. I will try to answer them all, but first and foremost let us be clear on one thing:

Organizations do value physical asset management. Physical assets are what bring value to an organization; there is no question about that. It is just that organizations either are not taking the right approach towards the physical asset management discipline, or they simply do not know what the right approach is. First and foremost it is very important for organizations to realize that things have to change and what worked yesterday does not work today and won't work tomorrow. Things must be done differently to achieve different results. Organizations need guidance to navigate through the paradigm shift and understand what it takes to overcome the challenges.

To explore the challenges that physical asset management is facing to fit into an organization, let us break the organization into layers and analyze the fit-in issue at every level: organizational, functional area and physical asset levels.

Organizational-Level Challenges

To demonstrate the challenge facing physical asset management at an organizational level, just ask anyone in your organization what "asset management" is about. I am sure you will hear all kind of different answers, some even very shocking. Mind you, it is no one's fault. It just happens that the words "asset management" have been used quite widely for a long time in maintenance, in finance, and in the consulting world. Now, with ageing physical assets in need of more care and more attention, the holistic physical asset management philosophy has taken center stage in many organizations globally. The use and interpretation

of the words "asset management" to define the holistic approach of managing physical assets has become a sensitive preoccupation for organizations and practitioners. It is still causing quite some confusion in many industries. You may have heard such things as:
- "Asset management—what does it do?"
- "Are we not already managing our assets?"
- "We've been practicing asset management for years. Why do we need an asset management function?"

In short, the holistic physical asset management approach is struggling to build its identity and, most importantly, to be recognized as a functional area in many organizations. In a sense, there are good reasons behind the quest for identity. We all appreciate, intuitively, that the practices of physical asset management have been around for many years. Maybe they were called differently in different industries, or maybe they were performed and shared under different functional areas. In some organizations, the practices may exist in the form of asset planning, capital planning, maintenance management, or in some cases asset management itself. You may walk into an organization and find that physical asset management practice is actually being deployed by the maintenance management function or in another by the capital planning function.

"So how does the physical asset management function differ from the other existing functions and how would they interact with each other?" you ask.

Good and valid question. To answer it, let us review and clarify some common confusions about the holistic physical asset management approach vis-à-vis other existing functions.

The most common confusion encountered in many organizations is between physical asset management and maintenance management as depicted in Fig. 4.1 below. Let us make it clear that:

$$AM \neq MM$$

where AM is Asset Management and MM is Maintenance Management.

Physical asset management and maintenance management are two distinct organizational functions with very different areas of focus. The main reason the confusion exists is that, in organizations where there is no physical asset management function, the maintenance management function would usually assume that role. Both functions deal with physical asset lifecycle delivery activities and must work together to address existing gaps and overlaps in organizations. In Chapter 6, we will analyze in more detail the gaps and overlaps between these two functions and look into how they should work together.

Fig. 4.1 "Who should I go to for this work request"

There is another, related and contentious debate:

Is MM⊂AM? That is, is Maintenance Management a subset of Asset Management?

While the two functions are totally different, this question frequently comes up. Technically, the answer is yes. Maintenance management involves managing physical assets for the short term while asset management involves managing physical assets for the long term. Now, by no means does this imply that the maintenance management function should report to the physical asset management function. What this means is that one has a top-down approach and the other has a bottom-up approach. Both approaches must be aligned, and both functions must work together to, meet the physical assets' technical service levels and, hence, achieve the organizational objectives. Whatever maintenance management does with the physical assets in the short term will have an impact on what physical asset management will do in the long term—more details on this topic in Chapter 6.

Let us move on to another burning question prevailing in many organizations:

Is PM⊂AM? That is, is Project Management (PM) a subset of Asset Management? For sure, most if not all of you will say a big NO. And I cannot disagree with that, but where do we draw the line between the two functions? Both the project management function and the asset management function must collaborate and work very closely together to deliver the right physical assets and meet operational requirements. Should project management deliver what physical asset management prescribes or should physical asset management accept what project management delivers? We will discuss this in more detail in Chapter 8.

"It looks like there is quite some confusion in how the two functional areas of maintenance management and project management differ from and interact with the physical asset management function," you observe. "So how are organizations currently managing their physical assets and how is that different to the holistic approach?"

Well, the major difference in the physical asset management approach from an organizational perspective can be uncovered when we compare how the public and private sectors manage their physical assets.

In the public sector, the physical asset management function is quite well set up, with well-documented physical asset management practices. For example, in many municipal sectors, asset management units or departments have been created, been fully staffed, and are relatively well run. In most of those organizations, either there is a fully developed and implemented asset management framework, with regularly produced asset management plans for different physical asset classes, or the process of creating this framework is underway. It is crucial to note that there are compelling reasons the physical asset management philosophy is so well developed and has a stable identity in the public sector.

The most important business drivers of the physical asset management function in the public sector are:
- To maintain required service levels as dictated by public needs, council, and regulators.
- To obtain capital funding to keep physical assets and infrastructure in a state of good repair.
- To be eligible for grants and subsidies from other levels of government.
- To remain in compliance with government requirements and applicable regulations, such as the Generally Accepted Accounting Principles (GAAP) to report on tangible capital assets.

On the other hand, in private sector organizations, responsibilities for physical asset management function fall under the maintenance or operations groups. Most of the time, you won't hear or see typical physical asset management terminologies, such as asset management framework or asset management plan, but the sound practices are there, well developed, and shared among different functions of the organization.

In the private sector, physical asset planning or capital planning are mainly dictated by the needs of the operations and marketing functions, and are implemented by the maintenance or engineering units, with support from procurement and finance. Here, most of the decisions related to physical assets and the capital budgets are made at operations and/or marketing levels, with main foci on new product development, new markets, and to improve maintenance and reliability of operations.

Organizational Silos

The difference in organizational approach to physical asset management is very much noticeable when we compare the functional areas prevailing in the public and private sectors.

Fig. 4.2 illustrates the typical functional area silos present in public sector organizations, where the functional areas are pretty much dispersed. In some cases, the functional areas have conflicting priorities, as demonstrated by the ToF discussed in Chapter 3. We have seen that, even though areas of focus of the functional areas can be similar or different, they eventually can blur the line of sight, if not managed properly. This results in functional paralysis and slow decision making, impacting physical assets performance and ultimately effectiveness of the whole organization.

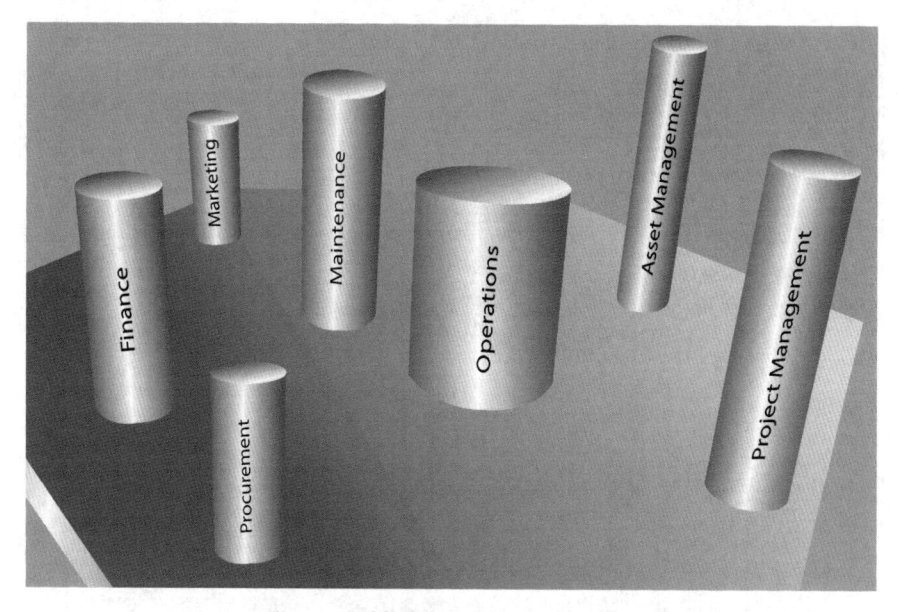

Fig. 4.2 Public Sector Functional Silos

Fig. 4.3 illustrates the typical situation in the private sector organization, where the operations functional area, working very closely with the marketing function, would control key functional areas closest to physical assets, such as maintenance, physical asset management (if it exists), and project management. In those organizations, decisions are made without a holistic view, which can sometimes result in a conflict of interest and poor or rushed decisions. It must be pointed out that many private organizations have reviewed/ are reviewing their strategies and are starting to take a holistic approach to improve performance and to gain competitive edge in their respective markets.

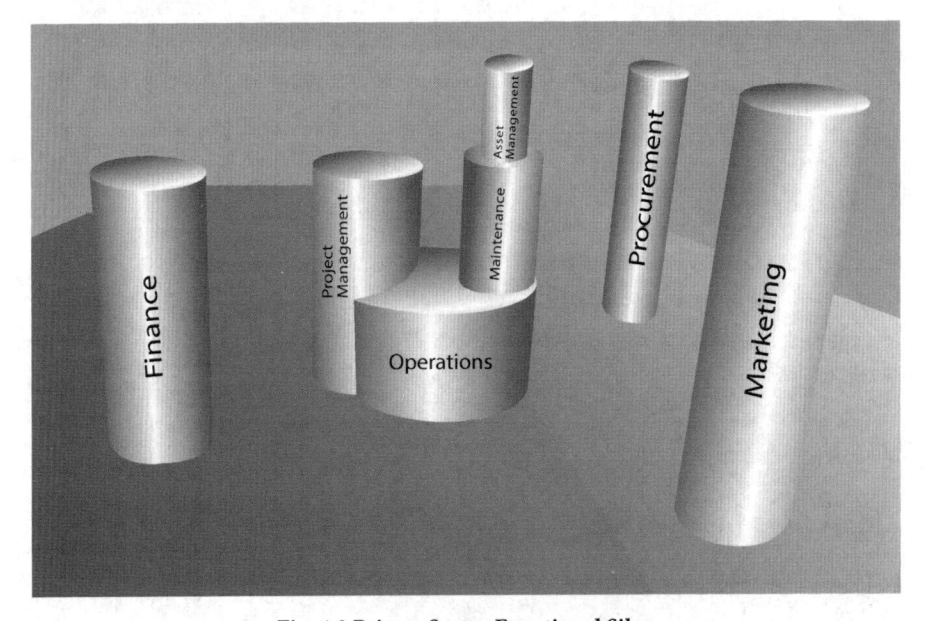

Fig. 4.3 Private Sector Functional Silos

The above diagrams illustrate the core of the challenge present in organizations today, with respect to physical asset management. What is crucial is how organizations structure themselves to deal with the existing functional silos and to streamline efforts from all functional areas to make the right decisions at the right time by the right people. We have the PAS55, the IIMM, or the ISO55000 (just to name a few) of the world to guide us through the process of implementing the physical asset management philosophy, but, at the end of the day, the buck stops at the organization level to decide on how physical asset management, as a function, fits into their structure and how interactions between the functional areas happen in a harmonious manner.

Case Study 4.1

It was late morning on a hot summer day. Jerry Detunam was in his office when he received a very rare call from the Vice-President of Operations, Robert Veristrik. Robert was enquiring about the power outage at the plant two days ago that shut down all operations for the whole day. An old transformer had caught fire. This was the second time in the last three months that there had been an equipment failure that had significantly disrupted operations. The first time it had been the failure of a switch gear in the main electrical panel.

"This type of plant downtime should not happen again. It caused significant loss to operations, and we received lots of customer complaints," Robert declared firmly.

Jerry took some time to respond to his VP Operations. Finally, in a soft voice, he said that he would look into the matter and get back to him with an action plan.

Deep inside, Jerry knew that the two main factors in both power outages were ageing physical assets and lack of maintenance.

Jerry Detunam had joined Mouroum Inc. as Asset Manager last year and reported to Robert Veristrik directly. The position of Asset Manager was a relatively new position in the organization, which also consisted of Operations, Maintenance, and Engineering Services.

Since arriving on the job, Jerry had been experiencing a few challenges:
- There was a total misconception of what asset management was all about in the organization.
- There was confusion surrounding his job, role, and responsibilities

- Jerry had been a bit confused about who he should have on his team.
- He had been encountering challenges in interacting and working with the other functional units.
- He was finding it hard to cope with the Operations group, who were making key physical asset decisions without any planning and business cases.

Jerry did not have the answers to all these questions. He did not know where to start or how to build the team he needed to provide the right physical asset management support to the organization. He was having a real challenge figuring out how to define his working relationship with the other functional areas.

There were questions trotting in Jerry's mind as he started thinking about how to respond to his boss. Questions like:

- How could he raise awareness about the holistic physical asset management approach?
- How could he get everyone to buy into the physical asset management philosophy?

Functional-Area-Level Challenges

At the functional area level, the challenges physical asset management is facing are quite complex. They are based on the interactions of the functional areas. As previously described, organizations have different functional areas, structured in different ways depending on the type and needs of the organization. In Chapter 2, if you recall, we looked at the different functional areas encountered in a typical organization. As a refresher, these functional areas are listed below:

Operations	Finance
Marketing	Procurement
Accounting	Human Resources
Maintenance	Engineering

Project Management					Information Technology (IT)

It is inconceivable, in today's business environment, to find an organization that deals with physical assets yet is without a procurement function or a maintenance function. Each function contributes, in its own way, to help the organization achieve its strategic goals, irrespective of how the organization is designed and structured. What is important is that all physical assets are taken care of during each of their lifecycle phases.

As a reminder, in Chapter 1, we described the six physical asset lifecycle phases: planning, acquisition, installation/commissioning, operation, maintenance, and disposal.

By now, we are all well aware that, for an organization to be successful, not only does it need to take care of its physical assets over their entire lifecycles, but it also must do so at optimum cost, while minimizing risks and providing the right level of service. Let us consider cost, for example. Fig. 4.4 below shows how costs vary during a physical asset lifecycle.

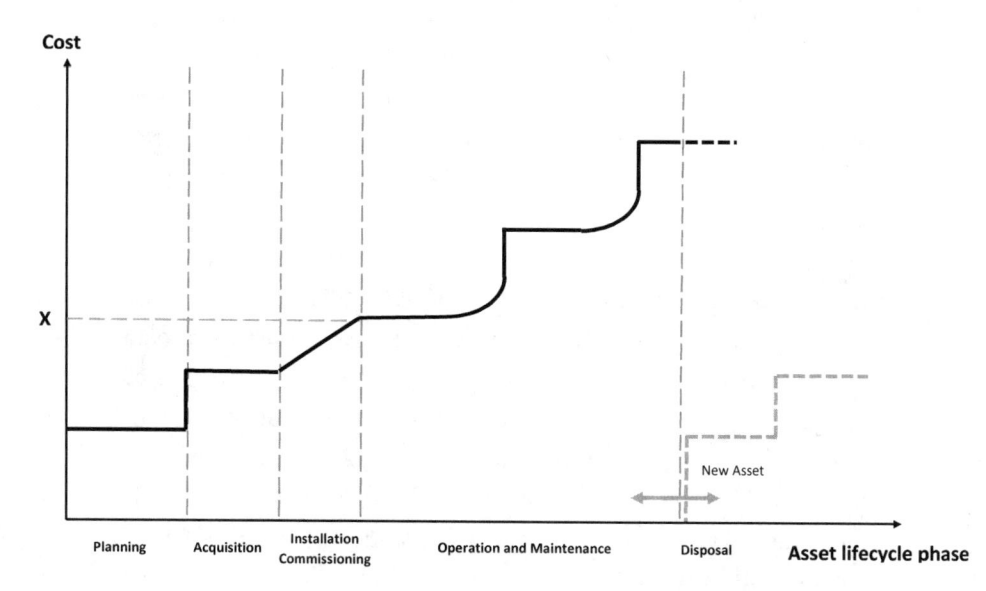

Fig. 4.4 Physical Asset Lifecycle Phases and Cost

During the planning phase, expenses are incurred to carry out needs assessment and studies. Further cost is incurred during the acquisition phase, and then continuously rises as we enter the installation/commissioning phase. By then, the physical asset would have cost the organization X amount of dollars, as shown above. When the physical asset is put into operation, it incurs operating and maintenance costs, which include periodical major rehabilitation costs. At some later point in time, the decision will be made to replace the physical asset which will involve a disposal cost.

During this whole process, each of the functional areas of the organization plays very important roles in the different physical asset lifecycle phases to influence cost, either while spending money or while monitoring and controlling the costs incurred.

"But not all functional areas are concerned with cost, are they?" you question.

You are correct. In fact, as seen in Chapter 3, the functional areas have many similar and different areas of focus and not all of them have "cost" as one. Even though a functional area may have an impact on cost, cost may not be one of its primary areas of focus.

Now, let us shift gears and examine the dynamics of the functional areas with respect to the physical asset lifecycle phases. Over time, as physical assets go through their different lifecycle phases, different functional areas interact with them. For example, operations will use a physical asset to produce goods or deliver services and, at some point in time, requests it to be replaced for certain reasons. At that point, the procurement department gets involved and goes through the tendering process to acquire the new physical asset. Then the project management or engineering team proceeds to install/commission and deliver the physical asset to operations. When operations have trouble with the physical asset, they call maintenance to repair and restore it to its original state. The finance department will most probably come every year to ask for evaluation of the physical asset and its remaining life so that they can depreciate it and update their records.

Isn't this a common scenario?

"Yes, this is what actually takes place in organizations," you quickly reply. "But what is the issue?" you continue.

The point I am trying to make is that the physical asset is like a "hot potato," being tossed around with no formal ownership or formal decision-making process. In many cases, decisions are made not from an organization's needs, but from a wants perspective, coming from different business units or functional areas, as they see fit to fulfill their specific mandates. Put yourself in the place of that physical asset (I know it is not easy to do — but try).

- Do you (physical asset) have a clear path of what will happen to you?
- Who will do what to you and when?
- Who should you go to or speak to if you have any problems?

In short, you have no clue about your fate in your whole lifecycle! Are they going to extend your life or are they going to get rid of you? Based on what criteria?

"Hmmm, that's an interesting perspective," you answer. "Too bad our physical assets can't do that."

So, on one side we have the physical asset lifecycle phases and, on the other side, we have the different organizational functional areas required to oversee those physical assets without clear ownership. In Fig. 4.5 below, we try to map the functional areas to the physical asset lifecycle phases to demonstrate the chaos that could be present in the life of a physical asset.

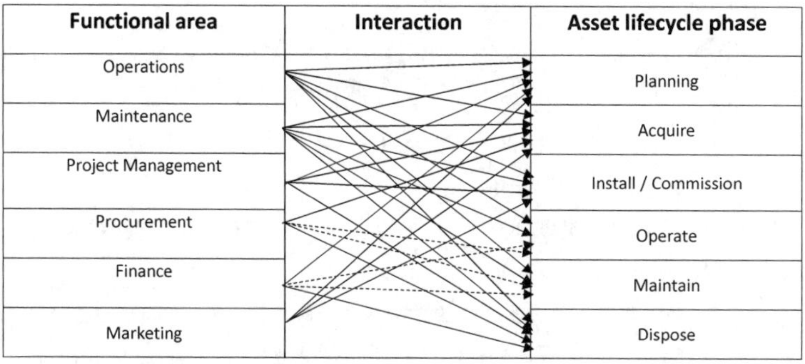

Functional area	Interaction	Asset lifecycle phase
Operations		Planning
Maintenance		Acquire
Project Management		Install / Commission
Procurement		Operate
Finance		Maintain
Marketing		Dispose

Fig. 4.5 Interaction of Functional Areas with Physical Asset Lifecycle Phases

Here is a challenge for you: try to count how many arrows point to the asset lifecycle column.

The figure above shows how chaotic the life of a physical asset can actually be. During the lifecycle of a physical asset, there is in fact a lot of interactions involving the different functions of an organization. These interactions form the core of the holistic physical asset management approach. Good or bad, in many organizations nowadays, physical asset management is carried out this way; i.e., by different functional areas for different purposes and (many times) in silos. As we already know, those different functional areas have different mandates and focus, and sometimes these different mandates can be detrimental to managing physical assets in the most effective manner.

To make things a little more complicated, those functional interactions with physical assets may take place at irregular frequencies—sometimes weeks, months, or even years apart depending on the type of physical assets or physical asset classes. The real challenge lies in the synergies of those functional interactions, the lack of which reflects ineffective physical asset management practices in many organizations, resulting in asset failures, downtime, poor decision making, and poor planning. How many times have we heard one functional area complaining about another functional area for not doing the right thing or for not providing the information required on a physical asset? I am sure you must have encountered at least one of the situations below:

Maintenance to Project Management: *"We were never consulted when you were selecting/designing this compressor!"*

Operations to Maintenance: *"Your staff just 'fixed' the conveyor and it broke again!"*

Finance to Maintenance: *"The building was just refurbished—I need the new valuation and life expectancy."*

At this point, I want to take a pause and remind readers that we are not trying to paint a poor picture of organizations vis-à-vis physical asset management, nor are we rating down the different functional areas dealing with physical assets. We all have to recognize that this is the reality of the situation prevailing in organizations. Having said that, we also recognize there are lots of organizations out there doing great things in the realm of physical asset management. There are great professionals out there performing excellent jobs in the different functional areas and things are working out relatively well. Maybe, in other organizations, some functional areas are still going through the transition period, which may have been triggered by changing expectations in their respective industries; for example, the adoption of physical asset management in the public sector to obtain funding from other levels of government, or the transition from maintenance to reliability in many corporations to comply with new and tighter regulations.

Physical Asset-Level Challenges

The challenges physical asset management is facing at the physical asset level boil down to the type of physical assets, themselves. If you look closely at the public and the private sector, you will notice each deals with a predominant type of physical asset. In the public sector, you will encounter mostly physical assets of the infrastructure-type (Chapter 2, infrastructure-centric organizations) requiring relatively little maintenance. These are **low-intensity maintenance** physical assets with longer economic life, such as roads, bridges, buildings, etc.

For example, a typical bridge will require intermittent maintenance, but will require mostly condition assessment and regular rehabilitation. The bridge itself will last over 100 years. Of course, if you look at the water or waste water area in the public sector, you will encounter *some exceptions* such as pumps, compressors, etc. But these are relatively few in comparison with the number of low-intensity maintenance physical assets being managed.

On the other hand, in private sector industries such as manufacturing, oil & gas, or utilities, you will encounter more **high-intensity**

maintenance physical assets of the equipment-type (Chapter 2, equipment-centric organizations), such as conveyors, blowers, mixers, and grinders, compared to low-intensity maintenance physical assets. This fundamental difference in a physical asset mix in the public and private sectors or in infrastructure-centric and equipment-centric organizations has a huge impact on the way holistic physical asset management is approached and the subsequent strategies for physical asset lifecycle delivery activities.

To illustrate the above statement, let us compare two physical assets: a bridge and a centrifugal pump as shown in Fig. 4.6 below. The bridge is a low-intensity maintenance physical asset, with lifecycle delivery activities ranging from yearly to ten-year frequencies. The centrifugal pump is a high-intensity maintenance physical asset, with lifecycle delivery activities ranging from daily inspections to yearly interventions. The bridge has an expected service life of over 100 years, while the centrifugal pump may have an expected service life of fewer than 25 years.

Physical Asset	Low-intensity maintenance BRIDGE	Freq.	High-intensity maintenance CENTRIFUGAL PUMP	Freq.
Expected Life	> 100 years		< 25 years	
Potential Maintenance Activities	Bridge Inspection	1 year	Inspect & clean filter	1 week
	Bridge superstructure deck cleaning	2 years	Check oil level & sign of oil leak	1 week
	Clean draining system	2 years	Exercise valve	1 month
	Lubricate bearings	4 years	Carry out vibration analysis	3 months
	Seal deck & fill cracks & joints	4 years	Check impeller clearance	6 months
	Replace seal joints	10 years	Carry out oil analysis	6 months
	Replace wearing surface	10 years	Check electric motor	1 year
			Check alignment	1 year

Fig. 4.6 Low-Intensity Maintenance and High-Intensity Maintenance Physical Assets

"What does all this mean from a physical asset management perspective?" you enquire.

Well, organizations with low-intensity maintenance physical assets focus more on long-term lifecycle delivery activities and condition assessment for better planning. On the other hand, organizations with high-intensity maintenance physical assets concentrate on short-term lifecycle delivery activities to meet required service levels. In both cases, the approach is different because of the different physical asset types involved. Organizations with a majority of one type of physical assets will have a different approach for how to handle those physical assets. Physical asset mix may shape up the organizational structure, the functional areas, as well as the skill sets required for the organization to manage its physical assets. For example, in the case of the bridge versus pump comparison: one functional area will have mostly engineering technicians and inspectors while the other will have mainly tradesmen, such as millwrights and electricians.

"At this point, we have seen the challenges faced by physical asset management at three different levels of the organization: organizational, functional area and physical asset levels," you observe. "What can we do about this, and what's next?"

Let me think for a moment. We know that, at the physical asset level, not much can be done about the challenges, because the type of physical assets is a given and organizations will have to address the needs accordingly. Now, to address the challenges at the organizational level would probably require major restructuring and culture change. This would take lots of time and effort and, in the end, it might prove to be a quite daunting and difficult task, but not impossible, to tackle the situation at that level. At the functional area level, there seem to be more prospects for tackling the challenges and achieving some level of success.

We saw earlier that physical assets go through a cradle-to-grave lifecycle with six distinct phases. We also saw that organizations have functional areas in place to take care of those physical assets at the different phases of their lifecycles, in order to realize the best value from them. This whole process—from the beginning to the end of a physical asset's life—might take many, many years and, in the process, many functional areas will be involved at different point in time. It has been clearly noted that the challenge occurs when these functional areas do not function in harmony and create functional silos each with their own specific mandates and goals. Gillian Tett put it brilliantly in her book, *The Silo Effect*:

> Silos can be useful but also dangerous. They have the power to collapse companies and destabilize financial markets, yet they still dominate the workplace. They blind and confuse us, often making modern institutions collectively act in risky, silly, and even stupid ways.

As mentioned earlier, organizations in the public and private sectors need functional silos to operate and deal with the complex nature of their business. But, when not managed properly, these functional silos can lead "to bureaucratic infighting, wasted resources, communication failures, and information bottlenecks, resulting in tunnel vision and mental blindness" (Source: Tett Gillian, The Silo Effect, 2015). This can sometimes cause damage, with dangerous and costly risks. Vivid examples close to physical asset management include the 2010 BP oil spill in the Gulf of Mexico and the 2014 General Motors crisis.

The big questions here are:

What can we do about these functional silos? Is there anything we can do to master them before they master us?

There is nothing wrong in work specialization and/or having functional silos in organizations, but wouldn't it be great if they all can dovetail nicely into each other, support one another, and all work together to achieve the same organizational goals? Wouldn't it be nice to have the silos represented by perfectly vertical cylindrical gears instead of plain cylinders? These cylindrical gears would mesh perfectly with one another, creating **harmonized silos** in the organization, as shown in Fig. 4.7 below.

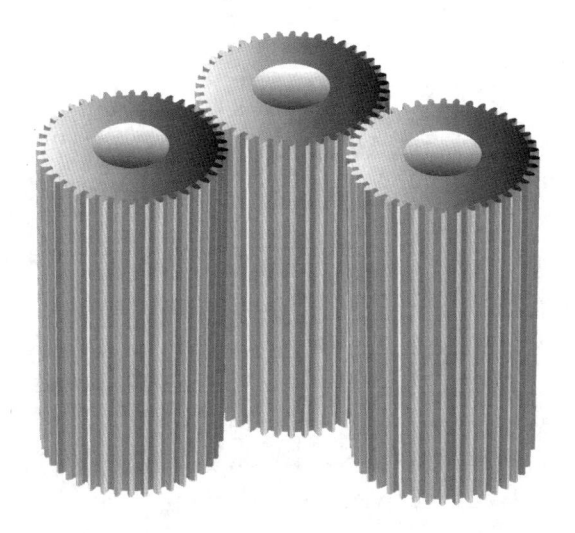

Fig. 4.7 Harmonized Silos

"But this only looks nice on paper," you quickly interject. "How do we make that happen in reality? What would the 'harmonized silos' structure look like in the physical asset management context?"

I have to admit that your excellent questions push me to think harder every time. You were asking how we can recreate those harmonized silos in an organization.

Well, there are two things that can happen with organizational silos:
1. Either they are too far apart to interact and communicate
2. Or they are too close to each other to the extent of interfering with each other.

"How about if the silos are not parallel to each other?" you ask.

That's a very good question. First of all the gears of the silos will not mesh perfectly and will create problem. In organizations this can translate into misalignment of certain functional areas to work towards the organizational objectives or poor line of sight. This may be because of poor direction or simply the wrong people in the wrong job. But let's not go there. Ok?

"Ok," you reluctantly agree. "But what if the silos rotate in different/wrong directions?"

Another good question, but you are getting ahead of the game. I will talk about that in Chapter 11 where we look at functional areas collaboration, harmonization and synchronization.

Going back to the above two scenarios we typically observe in organizations—some functional areas have poor/conflicting interaction with other functional areas or there is no interaction at all between the functional areas. No interaction/poor communication will translate into **gaps**, and poor/confliction interaction will translate into **overlaps**. In a nutshell, to create harmonized silos in an organization, we must address both the gaps between the functional silos and the overlaps between the functional silos.

For those of you who enjoy graphical explanation, in Fig. 4.8 below we map the level of interaction of different functional areas involved in physical asset management with the lifecycle phases of a physical asset. Looking along the vertical axis is the degree of interaction of each functional area in each lifecycle phase.

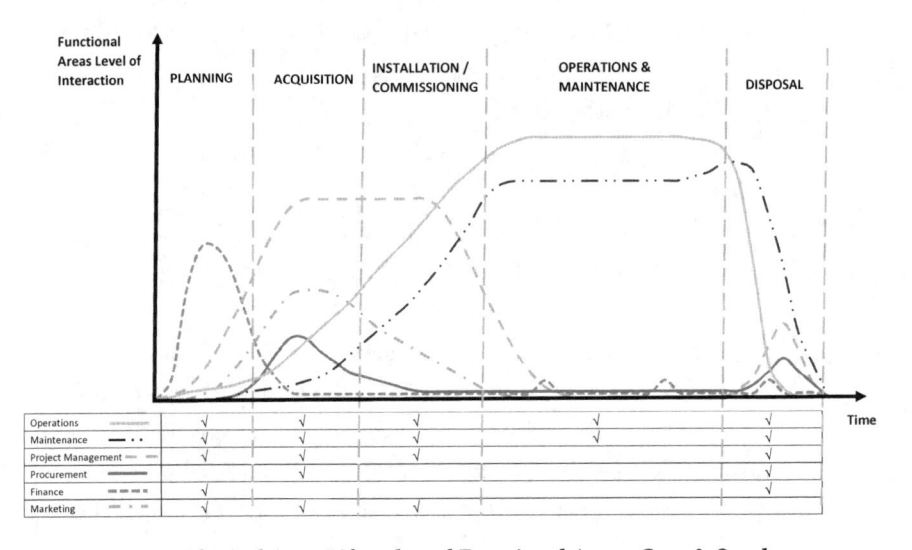

Fig. 4.8 Physical Asset Lifecycle and Functional Areas, Gaps & Overlaps

For example we can see the high level of involvement of Finance in the planning phase and lower level of involvement in the acquisition and disposal phase. Operations on the other hand show a gradual increase of involvement from the planning phase thorough acquisition and installation/commissioning. Involvement is at its peak in the Operations & Maintenance phase and declining in the disposal phase. Maintenance follows a similar pattern than Operations, except there is a little more involvement in the disposal phase. Project Management on the other hand is heavily involved in the acquisition and installation/commissioning phases.

Along the horizontal axis the diagram shows the timing of the involvement of different functional areas in the asset lifecycle phases. The combination of degree of interaction and timing of involvement of each functional area determines the potential areas of gaps and overlaps (G&O) that could exist between different functional areas in each of the physical asset lifecycle phases. Depending on the complexity and structure of organizations, the G&O between functional areas can fluctuate. Some may have more gaps than overlaps, or vice versa, depending on their timing of involvement and degree of interaction. This creates different dynamics among the functional areas in the lifecycle of the physical asset, that could ultimately affect the effectiveness of the holistic physical asset management philosophy.

To illustrate the G&O between functional areas, let us review the same examples that we looked at earlier:
- **Maintenance to Project Management**: *"We were never consulted when you were selecting/designing this compressor!"* The existing gap is that sometimes physical assets are selected or designed without considering maintainability and reliability. This is represented by the space between the two curves in the acquisition phase, where maintenance is lowly involved.
- **Operations to Maintenance**: *"Your staff just 'fixed' the conveyor and it broke again!"* The gap here could be that the

physical asset cannot perform at the level operations wants it to, no matter how much maintenance is done. In this case, even though maintenance and operations are heavily involved in the Operations & Maintenance phase, there may be potential misunderstanding or miscommunication.
- **Finance to Maintenance**: *"The building was just refurbished—I need the new valuation and life expectancy."* Here there is potential for overlap, whereby finance may have been providing different valuation numbers based on old valuation records.

Before we get to discuss what the actual physical asset management function should look like and how to address the existing G&O between the functional areas, let us analyze those functional areas one by one. Let us look at what their main mandates are in the context of physical asset management and review the existing G&O. Then we can determine how the actual physical asset management function can help to harmonize the interactions between the functional areas *by bridging the gaps* and *ironing out the overlaps* in order to achieve an effective physical asset management approach. In the remaining chapters we will be referencing Fig. 4.8 very often. It will be a good idea to understand it well and bookmark it.

CHAPTER 5 —
Operations

"We are what we repeatedly do.
Excellence, then, is not an act, but a habit,"
~ Aristotle

In Chapter 3, we mentioned that an organization consists of main functions and supporting functions that aid it to stay effectively in business and meet its strategic objectives. One of the key, if not the main functional area, of an organization is operations. Operations is that functional area that plans, organizes, coordinates, and controls the resources needed to produce goods and deliver services. It manages the traditional five Ps — Product, Plant, Process, Program, and People. Operations is the core function of any business, irrespective of its size or the nature of its business, and whether it is producing goods or delivering services. It is, in fact, the *raison d'être* of any organization, be it in the public or private sector or a non-profit entity. As a functional area, the main role of operations is to transform an organization's inputs into the finished goods or services. As shown in Fig. 5.1 below inputs include buildings, equipment, labour, raw materials, technology, and information while outputs are the actual goods and services.

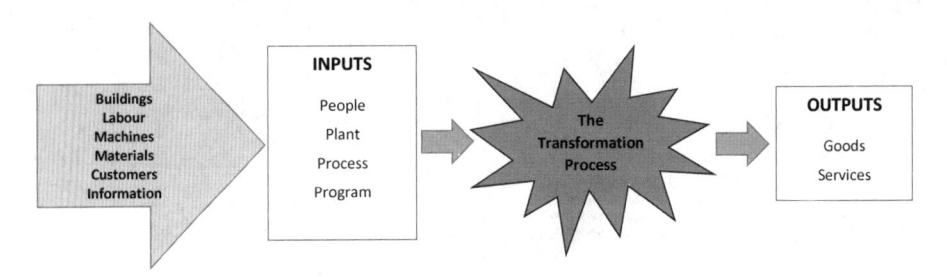

Fig. 5.1 Operations Transformation Process

The essence of the operations function during the transformation process is to add value. Value can be added at the strategic and tactical levels of organizations via the different elements of operations, such as:

Product design	Process selection
Supply chain management	Total Quality Management (TQM)
JIT and Lean systems	Forecasting
Capacity planning	Facility layout
Inventory and resource planning	Operations scheduling

Physical assets play a very important role in the transformation process, both in, manufacturing and in service organizations. For example, in manufacturing organizations where a tangible output is produced, well-maintained buildings, high-performance processing equipment, and reliable machinery are essential. In the service organizations, where an intangible output is delivered, such as collection of garbage, well-maintained collection trucks, safe landfills, and reliable compactors are required. Manufacturing and service delivery operations depend heavily on their physical assets to meet current and future demands and expectations. More and more stringent requirements are imposed on organizations from regulatory bodies, customers' demands, global economy constraints,

energy preservation, and climate change. This has resulted in a need for better physical asset stewardship, on three key fronts:
- Higher availability and reliability
- Financial sustainability
- Social responsibility

The operations functional area faces many challenges when it comes to managing physical assets on the above three fronts. This creates an operational dilemma, with conflicting interests, which many manufacturing and service organizations are facing today. Ultimately, the expectation is that physical assets must operate safely and provide maximum availability & reliability at optimum costs.

"But these are fair expectations," you claim.

Yes, I agree with you that these are fair expectations, but we must realize that they are not as simple as this to achieve. Do you want to see the proof? Well, look around your organization: that is the main reason we—operations managers, maintenance managers, finance managers, marketing managers, all staff, and all readers of this book—are hired. Our mandate is to find the right balance for the organization on all three fronts, to find the right formula to deal with the prevailing conflicts of interest, when it comes to physical asset management.

Let me elaborate further, with the following mind-teasing questions:
- Is it acceptable for this mining company to significantly reduce the maintenance shutdown period to give more production time to operations?
- Is it viable for this drinking water treatment plant to invest hundreds of thousands of dollars in a reliability-centered maintenance analysis on its process?
- Is it acceptable for this oil rig platform to use substandard parts on its critical physical assets?

At first glance, these are relatively simple questions to answer, but in many organizations, incorrect decisions are sometimes made, due to conflicting interests, for the sake of operations and the business at large.

To achieve the right balance amongst reliability, financial sustainability, and social responsibility, the operations functional area must work closely and interact with other functional areas in the organization. Together, they can take the right decisions and make sure the requirements of the operations strategy and the organization, in general, are met. As shown below in Fig. 5.2 organizations set their strategy, objectives and goals to achieve a certain level of service and performance. Operations, with the help and support of all other functional areas ensure they realize the best value from the physical assets in the most effective manner.

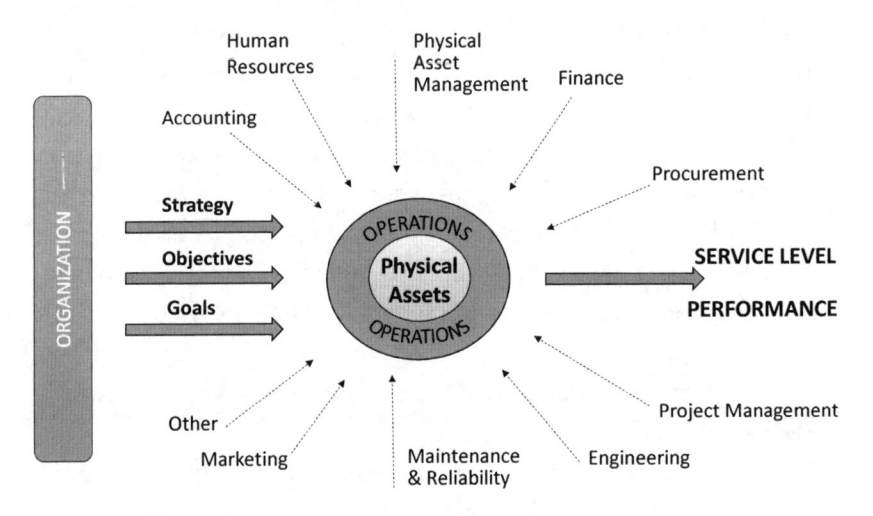

Fig. 5.2 Operations, Physical Assets, and Functional Areas

Below are some examples to illustrate:
- Operations will work closely with the marketing function to ensure future demand forecasts are met.
- Operations will work with engineering and project management to ensure that physical assets have the required short-term and long-term capacity.

- Operations will work with maintenance to ensure that the maximum capacity is always available when needed, to maximize throughput and achieve the best return on investment.
- Operations will work in harmony with the finance functional area to ensure capital funds are available for asset refurbishment or replacement in the future capital plan.

There are many such examples in an organization, where close interaction between functional areas is required.

Let us refer back to Fig. 4.8 for a moment. We see that the operations function is omnipresent throughout all phases of a physical asset lifecycle, working and interacting with other functional areas. This is no surprise at all because we very well know that operations is the core and the *raison d'être* of an organization. This does not mean that all other functional areas have to cave in to the operations function's needs and demands. There should be harmonized effort from all functional areas to support the operations function, by making the right decision for the organization with respect to physical assets.

Unfortunately, in the majority of organizations, this harmonized effort is not always fully present. What we may witness is some misalignment and miscommunication between the operations function and the other functional areas. This can result in the creation of G&O between functional areas, which ultimately affect the overall organizational effectiveness.

To illustrate this, let us look at the case study below.

Case Study 5.1

Demand for the goods produced by Bomlipi Manufacturing Ltd. has increased substantially over the last few months, and marketing is forecasting more demands going forward. In order to meet the increased market demand, operations had added a second production shift and, at the same time, they had increased the throughput of the production line from 100 kg/day to 120 kg/day. For the next couple of weeks, production was able to meet their daily target, even though the maintenance function had to be on the alert to attend some unusual, minor line stoppages. However, after another few weeks, the plant began to experience more frequent and more serious plant breakdowns.

Maintenance and reliability staff investigated the plant outages and equipment failures, and found that, in some cases, equipment were failing because of wear and tear and lack of maintenance. In other cases, they found that there were quite a few power outages due to the electrical system tripping on overload. The boiler used to provide steam to the production line was also unable to supply the required rate, resulting in the production line having to be stopped several times each shift. Maintenance, reliability, and engineering staff were still figuring out how they would tackle those plant breakdowns when plant operations sent an email informing personnel that operations had received instruction from senior management to increase production from 120 kg/day to 150 kg/day, due to more than forecasted increase in demands. This shift in production had to happen in three months' time, and Maintenance, reliability, and engineering staff were invited to attend a meeting the following week to discuss the situation.

The above case study depicts one of the realities many organizations face now and then. Because of shortcomings in planning and communication, and for the sake of short-term gains, physical assets are often put under tremendous stress, which affects their overall performance to deliver and reduces their economic life. In the long run, this can prove to be very costly for operations and a risky gamble for the organization.

Besides higher availability, reliability and social responsibility, organizations need to show financial sustainability to remain competitive. This statement applies not only to the private sector, but also to the public sector, where spending is being more and more scrutinized, where taxpayers are demanding more for less, and where the danger of privatization is looming. The onus is on the operations functional area to make sound decisions to improve productivity and show profitability, while managing risks at all levels. Basically, this means doing more and more with less and less. This can be achieved by optimizing costs incurred on the inputs (fixed and variable costs) of the transformation process, and by maximizing the output (total revenue) from the transformation process. Fig. 5.3 below (which many of you are familiar with) illustrates the relationship between total cost and total revenue to maximize profitability.

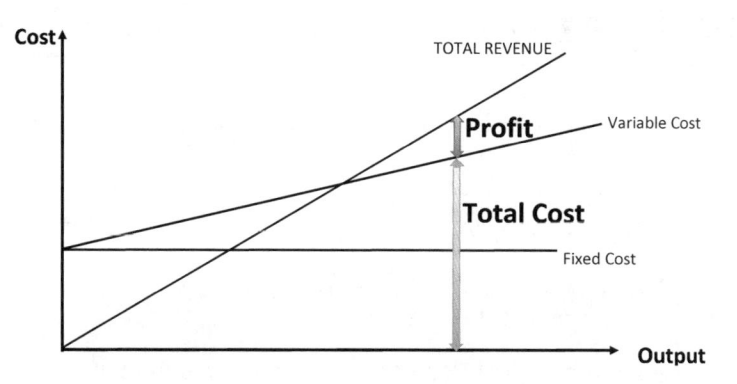

Fig. 5.3 Operating Costs and Maximizing Profit

It is vital for operations to measure how well it is using its resources to produce the required goods and services. Productivity is a good measure to show how effective the transformation process is and is the ratio of output to input.

Productivity = Output/Input

Organizations measure productivity differently and at different levels depending on what they want to track performance on to identify areas for improvement. There are partial, multifactor and total productivity measures. From a physical asset perspective organizations use partial productivity measures such as maintenance productivity, energy productivity, capital productivity or inventory productivity.

Another measure commonly used in many organizations to demonstrate the effective use of physical assets during the transformation process is the **overall equipment effectiveness** (OEE). OEE is a productivity management metric that brings operations, maintenance, and engineering together to achieve higher level of performance from the physical assets. In a nutshell, OEE optimizes the utilization of inputs to generate maximum output at the right quality level.

OEE = Availability x Performance Efficiency x Quality Rate, Where,

> Availability = Runtime/Scheduled Time
>
> Performance Efficiency = Theoretical Cycle Time/Actual Cycle Time
>
> Quality Rate = # Good Units/# Total Units

To improve productivity and OEE, the operations function cannot do it on its own. It needs the total support and help from the other functional areas of the organization. If the maintenance management function does not do a good job in maintaining consistent availability, OEE will ultimately be affected. If project management/engineering do not get the physical asset/asset system design spot on it will impact OEE. By measuring and tracking OEE, the operations function can bring together the other functional areas' ToF and align them with operational needs. OEE is a great vehicle to tackle the G&O and to create harmonization with all functional areas to improve physical asset management.

"Why do we have the G&O between operations and the other functional areas, then?" you ask. "Are these G&Os preventing operations from functioning to its full potential?"

Yes, the G&O are without a doubt preventing the operations function from meeting its mandate. Rather than supporting operations, some functional areas may actually be creating more

hindrances. Those hindrances could be attributed either to the ToFs of the different functional areas, which are not fully aligned, or to the following operational requirements scenarios:
- Operational needs are not clearly defined in advance
- Operational needs are constantly changing
- Operational needs are not communicated in advance
- Operational needs are not addressed promptly
- Operational needs are misunderstood or disregarded

As a matter of fact, with the above challenging scenarios prevailing in organizations, many functional areas will undoubtedly make the wrong decisions and will fail to support operations. This has been clearly demonstrated in Case Study 5.1, where the maintenance, reliability and engineering groups were not in a position to fulfill the physical assets' performance requirements to meet changing operational needs. In effect, other functional areas may get caught in similar trap and may end up making the wrong decisions with physical assets.

For example, procurement may purchase the cheapest physical asset on the market if they do not know how it will impact quality. Project management may deliver the physical asset with a smaller than required designed capacity if they are not made aware of the short-term and long-term capacity requirements. Or finance will not provide the required funds to invest in the new technology required to boost production if they are not convinced on the return on investments

"So what?" you ask. "What does this have to do with physical asset management?"

Well, this is where the holistic asset management approach can support operations, and bridge the gaps or iron out some of the overlaps with other functional areas. This will help the organization, as a whole, to make the right decisions at the right time.

Holistic physical asset management can help the operations function by:
- Defining and documenting operational service levels

- Translating those operational service levels into physical asset service levels and functional specifications
- Defining current needs and identifying future needs via needs assessments
- Developing physical asset lifecycle delivery activities to maximize economic life
- Analyzing operational risks and consequences pertaining to ageing or low-performing physical assets
- Working closely with the maintenance function to increase physical asset reliability and to improve performance
- Working closely with engineering and project management functional areas to ensure the right physical assets, with the right performance specifications, are designed and acquired.
- Developing business cases and carrying out feasibility studies for capital projects or process improvement initiatives

Operations will hugely benefit from the support of the holistic physical asset management philosophy to fulfill its mandate as a functional area and to fulfill the short-term & long-term needs of the whole organization.

CHAPTER 6 —
Maintenance Management

"An ounce of prevention is worth a pound of cure."
~ Benjamin Franklin

When we talk about physical assets, the maintenance function is at the forefront of the discussion. The maintenance of physical assets plays a pivotal role in helping organizations achieve operational excellence and meet their strategic objectives. The main mandate of maintenance is to take care of physical assets and make them available to operations.

It should be noted early on that, in many organizations, when we talk about operations, it sometimes includes the maintenance management function.

Because this book deals mainly with physical assets, it is totally fair that the maintenance management function is reviewed as a separate function to clarify better its linkages with physical asset management.

By definition, we know that the purpose of ***maintenance*** is "to ensure that physical assets continue to do what their users want them to do." (John Moubray, RCM II Reliability-Centered Maintenance, Second Edition)

In organizations, the users are traditionally the operations functional area, whether in the manufacturing or service delivery sectors. Physical assets are built or acquired for operations to meet specific needs and requirements, based on a certain return on investment scenarios and a specific economic life. During the economic life of the physical assets, the maintenance function deals with ensuring that the physical assets:

- Can fulfill all functions as required by operations
- Are available to operations, as required
- Will last the designed economic life with optimum lifecycle cost
- Are safe and compliant at all times

In other words, physical assets are very critical for operations to meet its organizational objectives because the assets must fulfill the requirements levels listed above. And the maintenance function, to that effect, plays a key role to make all that happen.

A typical maintenance management function consists of important elements, such as:

- **Information database**, which includes a physical asset registry with documentation and drawings, and physical assets' history, such as queries, reports, and key performance indicators. Most of this would be managed (ideally) within a computerized maintenance management system (CMMS).
- **Maintenance strategies/tactics**, such as condition-based, preventive, or detective maintenance, which include the actual practices and procedures for each and every task of each physical asset.
- **Work management process**, which includes the process to manage work requests and work orders throughout their lifecycle. This entails procedures for how to plan, schedule, and coordinate the execution of different types of work using the CMMS, based on priority of the work and availability of resources.
- **Inventory management**, which consists of the identification, procurement, storage, and issuing of all parts related to the

maintenance tasks. It includes both the physical store room organization, as well as the procedures within the CMMS.
- **Personnel management**, which deals with the team structure, staffing with the right skill sets, training, and assessment of the workforce. It also involves the management of external contractors.
- **Supporting activities**, which entail all other activities that the maintenance function needs to provide with regards to physical assets such as health and safety, environmental, energy conservation, and other industry specific activities.

All the key elements of the maintenance management function are graphically shown in Fig. 6.1 below.

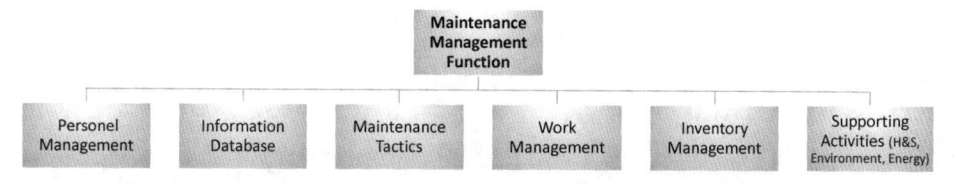

Fig. 6.1 Elements of the Maintenance Management Function

Via these elements, the maintenance management function has to execute different mandates so as to meet the physical assets' requirements of the organization. Below are some of the main mandates:
- To maintain up-to-date records of all physical assets in the CMMS system
- To ensure each physical asset has a set of maintenance practices and procedures in the CMMS
- To ensure all proactive maintenance tasks are carried out promptly
- To ensure all reactive maintenance tasks are carried out as required
- To document all works performed on physical assets in the CMMS
- To manage and control all parts inventory

- To track all maintenance costs and physical asset performance indicators
- To ensure compliance with all applicable legislations

"It looks like there is quite a lot to accomplish," you declare. "How does the maintenance management function go about fulfilling those mandates?"

First and foremost, the maintenance management function needs to do the right things, to the right physical assets, at the right time. This is very important to ensure physical assets are taken care of during their economic lives, when lots of wear and tear and random failures are expected to occur. Some of those failures can be prevented, some can be predicted, some can be detected, and others have to be dealt with differently.

"I believe this is the key responsibility of the maintenance management function," you say. "What approach can we take to deal with physical asset failures?"

Well, the approach should be proactive, as opposed to being reactive. No organizations like to deal with reactive situations, especially when physical asset failures have already occurred, with significant risks and consequences; however, not all failures can be foreseen, and there will always be some reactive situations. But organizations need to be prepared and ready to deal with those situations in a responsive manner and to be able to manage the risks and consequences. Fig. 6.2 illustrates the planned and unplanned maintenance strategies that the maintenance management function should have in place to manage physical asset failures.

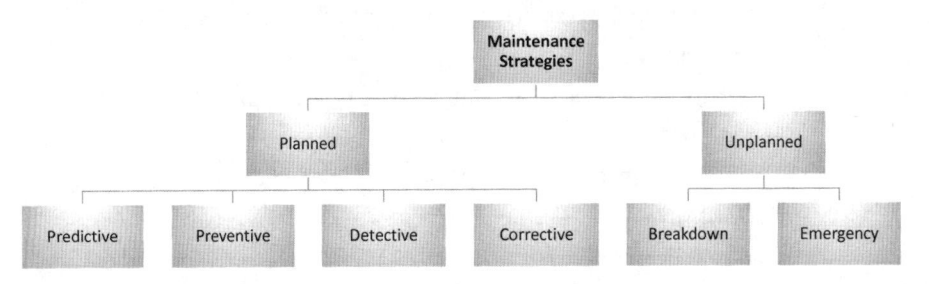

Fig. 6.2 Planned and Unplanned Maintenance Strategies

The planned maintenance strategy consists of predictive, preventive, detective, and corrective maintenance tactics. These are all activities to predict and prevent failures before they happen, so that the organization does not suffer the consequences of those failures. A good target for a maintenance management function is to achieve above 80% of planned maintenance. The other 20% or less will be dealt with by the unplanned maintenance strategies. These are situations where the maintenance management function reacts to emergencies and breakdowns.

"That 20% or less of emergencies and breakdowns can really hurt organizations, can't it?" you enquire.

In a sense, you are right. It can happen, especially if organizations have not done a thorough analysis of their critical physical assets and failure consequences. That is why you see lots of organizations/maintenance management functions evolving into reliability, which we will discuss in more detail in Chapter 7. Organizations and their maintenance management functions want to make sure they have all their planned *and*, for that matter, all their unplanned maintenance strategies spot on and under control.

"Wouldn't this be a normal approach for organizations to take with their physical assets?"

You bet. But this is more complicated than it looks on paper. On one side, you have all the physical assets with their maintenance tactics and, on the other side, you have the business needs with limited resources and other constraints. In order to achieve excellence in the maintenance management function, organizations need to invest in the right resources. Ultimately, the path to excellence revolves around executing the physical assets' maintenance tactics in the most effective and efficient manner as shown in Fig. 6.3 below.

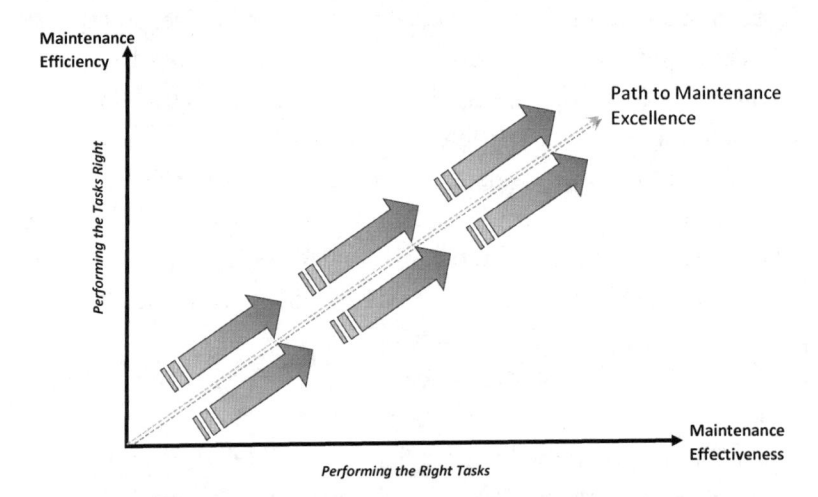

Fig. 6.3 Maintenance Management Path to Excellence

Maintenance effectiveness is built on good foundational maintenance practices to do the right tasks at the right time on the physical assets. Maintenance efficiency is built on good foundational maintenance processes to do the right tasks right to the physical assets. When effectiveness and efficiency are combined, there is the element of cost that comes into play. We are all very well aware that too little maintenance for physical assets is obviously not good, but too much maintenance is not good either. In Fig. 6.4 below, level of maintenance is plotted against cost. As the level of maintenance increases, the cost of maintenance goes up. At the same time, if we spend money on the right maintenance for the right physical assets, we will see the breakdowns and emergencies go down (i.e., the 20% reactive work decreases). We all know that reactive work, in general, costs many times more than proactive work and hence we see a sharp decrease in cost with more proactive work. But, like everything else, there is a balance that is reached where the cost of more maintenance will not bring much value to the organization.

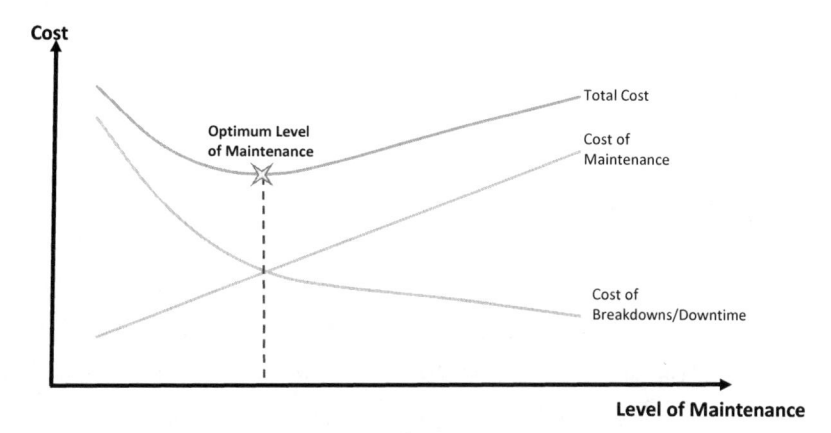

Fig. 6.4 Maintenance Cost Optimization

This balance point is called the *optimum level of maintenance*, and the maintenance management function must make sure to operate close to and around that balance point.

In Fig. 5.3 from Chapter 5, we saw that cost control is very important in relation to output and productivity. Besides taking care of the physical assets (by doing the right maintenance activities to sustain required service level and to minimize risk), the maintenance management function must also ensure cost optimization.

"Wait a second," you exclaim. "Isn't this the same definition for physical asset management?"

Yes, and rightly so. This is a similar statement that we encountered earlier in Chapter 3 for the definition of physical asset management. The truth is that the ToF of both the maintenance management and the physical asset management functions are very close, especially when it comes to cost. This is why, in many organizations, you will see the two functions either under the same umbrella or working closely together to support each other.

Referring back to Fig. 4.8, we see that the maintenance management function, similar to operations, is also omnipresent in the physical asset lifecycle, except in the planning stage, where it is lightly involved, depending on the circumstances and the organization. Again, there is no surprise here, since physical assets are the main focus of the maintenance management function. Maintenance has to interact closely with operations, because they are maintenance's main customer However, in many organizations, we witness the maintenance management function struggling to cope with and meet operations' demands and expectations. This is partly due to the intricacies of the operations function itself, as explained in Chapter 5, but it is also partly due to the lack of interaction between the maintenance management function and other functional areas over the entire lifecycle of physical assets. The result is G&O, and hence adverse impact on organizational effectiveness.

The maintenance management function may experience G&O with other functions at the asset planning stage, during acquisition, or even during installation or commissioning, where the maintenance management function is not always consulted or engaged early enough. I am sure many maintenance practitioners would agree with me on this one. As a result, the maintenance management function inherits physical assets they have to maintain, without having had the opportunity to have any/minimum say in the assets' design or configuration. On occasion, we may find G&O when the physical asset is in the operations phase. For example, the application of the right physical asset lifecycle delivery activities at the right time: who decides whether to refurbish or to replace the physical asset? When and why?

Case Study 6.1

The treatment plant at Santipisa Ltd. has three blowers to supply air to the aeration tank. During normal operations, two of the three blowers must be in service at all times to meet the demands of operations. The third blower is used as a back-up in case of failure of one of the duty blowers.

The blowers are all thirty years old and were installed when the plant started operation. The maintenance department has been doing a good job to keep them functioning up to now to meet operational demands. All the maintenance and repair data of the three blowers is recorded in the company's CMMS.

For the last few months, operations has been noticing a drop in the performance of the blowers, which was affecting the overall efficiency of operations. On top of that, they are forecasting an increase in demand in the coming years, which will require more capacity from the existing blowers. Operations have put in a request to engineering to replace the blowers over the next year and a half. The swap will be done one by one, so as not to disrupt operations.

Maintenance was not aware of the plan to replace the blowers. As was previously scheduled, they proceeded with extensive refurbishment on the three blowers over the next six months or so, spending a significant amount of money in the process.

This is a very common situation in many organizations where, many times, wrong or untimely decisions are made due to lack of communication and synergies between functional areas. G&O between the maintenance management function and other functional areas must be addressed to better manage physical assets throughout their lifecycles.

The physical asset management function can help organizations address those gaps and overlaps. Fig. 6.5 below shows some typical physical asset lifecycle delivery activities for the maintenance management and physical asset management functions, when they co-exist as two different functions in an organization. You will recognize in the bottom row the tactical activities for the maintenance management function, as discussed earlier in the chapter. The other lifecycle delivery activities in the top row (design modification, refurbishment and replacement) are usually the strategic activities covered under the physical asset management function. Note how the two dotted rectangular areas overlap with each other.

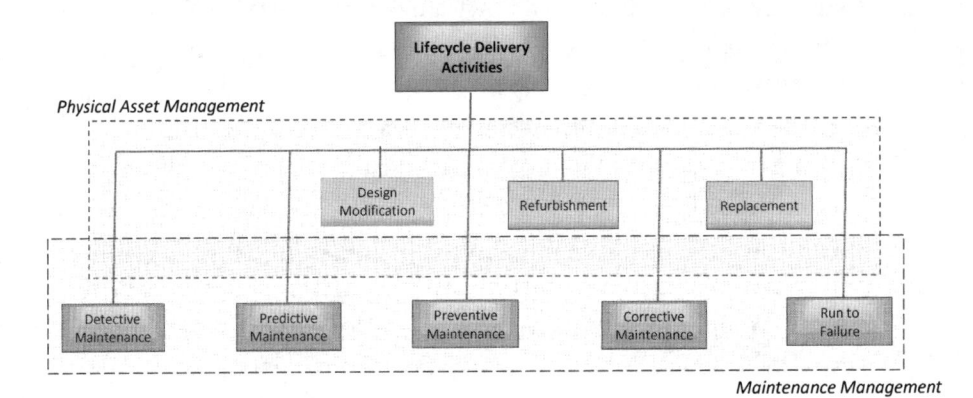

Fig. 6.5 Typical Asset Lifecycle Delivery Activities

The maintenance management and asset management functions complement each other, and we cannot emphasize enough the importance of harmonious interaction and communication between the two, throughout the entire life of physical assets.

Physical asset management can support the maintenance management function in many ways, such as:

- Introducing the reliability-centered philosophy early on, at the planning, design, acquisition, and commissioning stages
- Helping maintenance to understand and define physical asset performance standards and service levels
- Helping maintenance to develop and improve on physical assets' lifecycle delivery activities
- Helping maintenance to develop and analyze lifecycle cost from physical asset historical data in the CMMS
- Helping maintenance and operations make decisions regarding low-performance physical assets with repetitive failures, based on deterioration curves.
- Developing business cases and cost benefit analysis for maintenance to replace/refurbish/upgrade physical assets

A good way to demonstrate how physical asset management can improve the effectiveness of the maintenance function and the management of physical assets, is through the performance indicators. There are many maintenance management performance indicators that are used in many different industries to measure the maintenance function effectiveness. In Chapter 3, we showed that the ToF of the maintenance function is *cost-availability-reliability*. Physical asset management can improve those performance indicators, while supporting the maintenance management function as well as improving the way physical assets is managed.

As already mentioned, **cost** in maintenance must be measured and optimized to get the best returns. Activity-based costing helps to record the labour and materials used every time there is an intervention on a physical asset, to build the history in the CMMS. Physical asset management can use the cost history to translate these numbers into life cycle cost, analyze and forecast the trend(s), and help make decisions on the long-term strategy for the physical asset.

Availability is a key performance indicator for maintenance to track the effectiveness of the service level provided to operations. On the other hand, operations has to be satisfied with physical assets' availability achieved by maintenance to match full production schedule and maximize capacity. Consistent physical asset availability helps operations, and the organization in general, to better strategize and plan for future demands and capacity utilization, which means less pressure for more capital investment.

All physical assets are designed with a certain inherent **reliability** to satisfy specified performance standards under certain operating conditions. A poor design implies lower inherent reliability. The task of the maintenance management function is to preserve that inherent reliability under the specified operating conditions, by performing the right inspection and maintenance activities at the right frequency, as prescribed. Good work management processes, together with proper planning and scheduling practices, are essential to execute all the maintenance tactics, as required. In many industries, the maintenance management function measures the planned maintenance completion ratio to that effect. This shows how well maintenance is taking care of the physical assets to preserve inherent reliability and economic life.

The next chapter will take us deeper into the realm of inherent reliability and reliability engineering to show how they are applied in physical asset management.

CHAPTER 7 —
Reliability Engineering

*"Creativity isn't worth a thing if
it isn't served with an equal amount of reliability."*
~ Anton Peck

Let us start off with some definitions, to better understand what we are going to be dealing with in this chapter.

Reliability is defined as "the probability that an item will perform a required function without failure under stated conditions for a stated period of time." (Patrick D. T. O'Connor, Practical Reliability Engineering, Fourth Edition)

Reliability engineering is defined as "the application of engineering knowledge and specialist techniques to prevent or to reduce the likelihood or frequency of failures." (Patrick D. T. O'Connor, Practical Reliability Engineering, Fourth Edition)

"This is all good," you say. "But why is reliability not dealt with under the maintenance management function?"

In fact, the main reason for this is to draw to the attention of readers and physical asset management practitioners that there is a subtle difference between maintenance and reliability. The goals of this chapter are to clarify the differences between the two, to show

how each complements the other, and most importantly to reveal how they are related to physical asset management function.

Too often, in many organizations, we see that maintenance and reliability philosophies are mixed up or interchanged with one another. The result is that one loses its functional identity, focus, and importance at the expense of the other. Within an organization, both should co-exist and complement each other to take care of the physical assets. Again, that will depend on the needs and complexity of the organization's physical assets.

In Chapter 6, we saw that the definition of **maintenance** is "to ensure that physical assets continue to do what their users want them to do." (John Moubray, RCM II Reliability-Centered Maintenance, Second Edition). If we combine maintenance and reliability, we have reliability-centered maintenance. The definition of **reliability-centered maintenance** is "a process used to determine what must be done to ensure that any physical asset continues to do what its users want it to do in its present operating context." (John Moubray, RCM II Reliability-Centered Maintenance, Second Edition).

Notice the differences between the definitions of maintenance, reliability and reliability-centered maintenance: conditions, time, and operating context.

The reliability-centered maintenance (RCM) philosophy is a structured methodology that consists of seven foundational questions, as shown in Fig. 7.1, below.

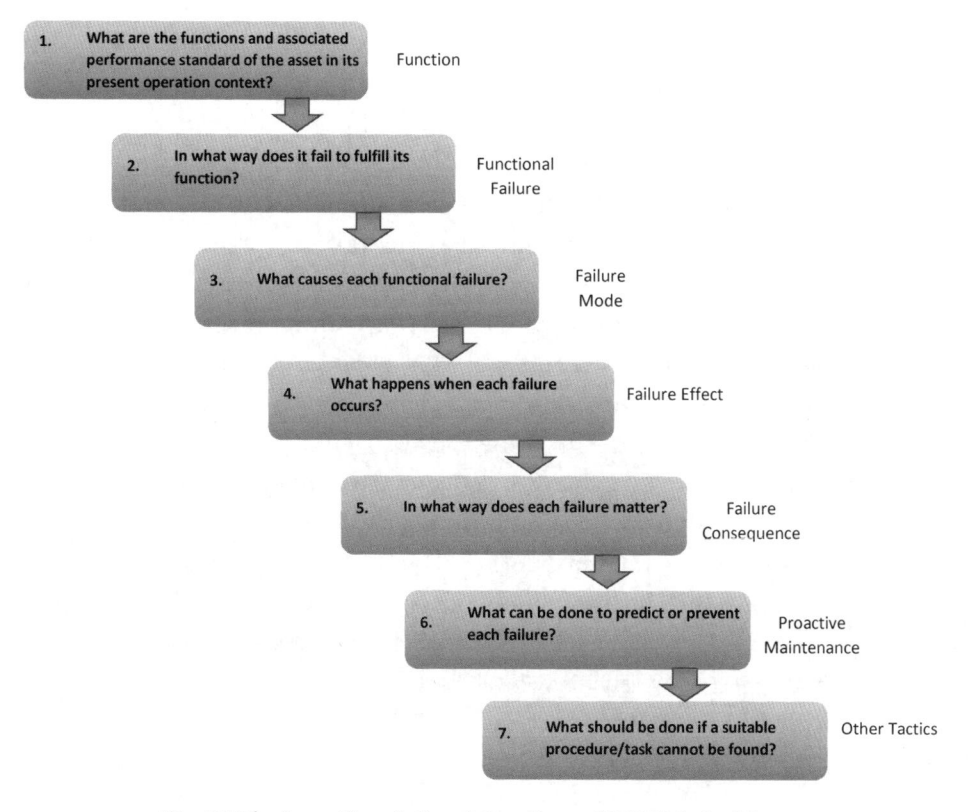

Fig. 7.1 The Seven Foundational Questions of RCM Methodology

1. What are the functions and associated performance standard of the asset in its present operation context? — Function
2. In what way does it fail to fulfill its function? — Functional Failure
3. What causes each functional failure? — Failure Mode
4. What happens when each failure occurs? — Failure Effect
5. In what way does each failure matter? — Failure Consequence
6. What can be done to predict or prevent each failure? — Proactive Maintenance
7. What should be done if a suitable procedure/task cannot be found? — Other Tactics

The RCM methodology was initially used in the aviation industry in the 1970s to determine the optimum maintenance requirements for aircraft and improve their reliability. The RCM methodology is a structured, proactive approach defined by the technical standard SAE JA1011, Evaluation Criteria for RCM Processes. It is used to analyze physical assets and asset systems, to assess their risk to the organization, to evaluate their consequences of failure, and ultimately to come up with strategies/tactics to manage the consequences of failure. The RCM methodology evolved from the Failure modes and effects analysis (FMEA) or Failure modes, effects and criticality analysis (FMECA), which is a step-by-step approach for identifying all possible failures in a design, a manufacturing or assembly process, or a product or service. FMEA was initiated in the 1940s by the U.S. military, and was further developed by the aerospace and automotive industries. Several industries maintain formal FMEA standards or apply the RCM methodology.

Other methodologies—such as Root Cause Failure Analysis (RCFA) and Failure Reporting Analysis and Corrective Action System (FRACAS) —are after-the-fact reactive approaches to improve the current reliability of physical assets and asset systems. In the reliability centered maintenance world, the focus is not to prevent failures of physical assets from happening, but is primarily to manage the consequences of those failures, while ensuring performance standards are achieved with minimum risks to the organization. Now, reliability is not confined to physical assets only. Physical assets do not bring value to the organization on their own. They bring value when they form part of a system and/or a process where reliability is critical. From an organizational and operational perspective, there are three levels of reliability that are important:

- Physical asset reliability
- System reliability
- Process reliability

At all times there must be a clear line of sight across all three levels of reliability to show how the process reliability is linked with the physical asset reliability via the system reliability. This will explain how physical asset performance standards are related to process/operational service levels. Functional areas, such as physical asset management, reliability engineering, and maintenance management must ensure that these service levels are all aligned and are attained through proper management of physical assets to achieve the organization's strategic goals.

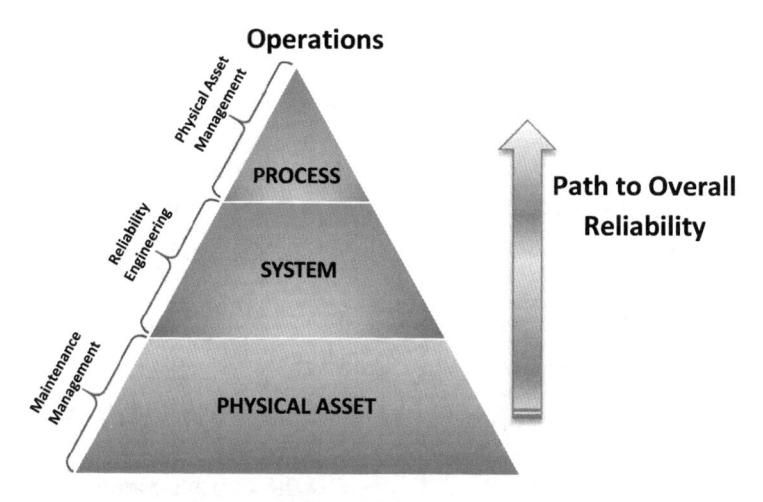

Fig. 7.2 Levels of Reliability in Operations

"How do the three functional areas make that happen?" you ask.

Well, think of the situation this way: As shown in Fig. 7.2 above the physical asset management function deals with the situation from the operational to the process service level; reliability engineering deals with the situation from the process to the system service level; and, maintenance management deals with the situation from the system to the physical asset service level.

As mentioned earlier, in some organizations, both the maintenance management function and the reliability engineering function are merged into one functional area, while other organizations keep these two functions separate. In either case, it is not a problem, as long as the functions are well aware of their mandates and the organizations understand their importance in physical asset management.

The main mandates of a reliability engineering function in an organization are to:
- Assess the criticality of physical assets/asset systems, based on the organization's operating context

- Develop maintenance strategies and tactics, based on failure mode effect analysis
- Identify root causes of failure, using fault tree analysis and RCFA
- Assess reliability of physical assets, asset systems and process

The reliability engineering function, whether embedded in the maintenance management function or as a standalone function, is very important to:

- Ensure the line of sight between the physical asset performance level and the operational service level
- Focus the asset lifecycle delivery activities on the most critical physical assets/asset systems.

Note that, in Fig. 4.8, we did not show the reliability engineering functional area because we assume it is embedded in the maintenance management functional area of the organization. It must be recognized that, in many organizations, this embedment can create gaps or even overlaps, which could jeopardize the effectiveness of managing physical assets. Below are some possible situations that can arise:

- No clearly defined and documented operating context
- No clearly defined and agreed-upon physical asset and asset system performance standards or service levels
- No clear line of sight between physical asset service level and operational service level
- Lack of thorough FMEA on critical physical assets/asset systems
- Lack of thorough RCFA and follow up for improvement
- No reliability requirements analysis on new physical assets and asset systems
- No review of redundant physical asset systems for optimum reliability

To illustrate some of the challenges mentioned above, let us review the case study, below.

Case Study 7.1

Tayerase Manufacturing has two pumps installed in its washing bay to supply hot water to the crate washing machine. During normal operation, only one pump is required to be in service. The second pump is used only when the duty pump fails or when it is undergoing maintenance.

Midway through the shift on a Tuesday, the duty pump failed. The maintenance manager knew that all maintenance activities on the duty pump were completed and on time. The standby pump was started, but it took one hour to get it going due to a faulty switch. Tayerase Manufacturing had to incur one hour of production downtime, after which normal production resumed with the standby pump.

On Friday of the same week, the standby pump failed while the duty pump was still under repair. Due to this failure, Tayerase Manufacturing's whole production line ground to a halt. The downtime lasted almost half a day, the time it took for the technicians to swap a component from the under-repair duty pump to the standby pump and get the standby pump up and running again.

So what can the reliability engineering function do to help Tayerase Manufacturing? And, how can physical asset management provide support?

The answer revolves around the concept of RAM—reliability, availability, and maintainability. How reliable was the system with a duty and a standby pump? What else can we do to improve reliability – a more reliable pump? Has maintenance been able to maintain the inherent reliability of the pump and the pumping system to achieve required availability? Has all proactive maintenance been carried out as specified? Were there any operations misuse or abuse? How maintainable is the pump itself and the whole pumping system? How easy it is to disassemble the pump or to replace components? Do the repairs take too long?

All these are questions that stem from the RAM concept as detailed in Fig. 7.3 below.

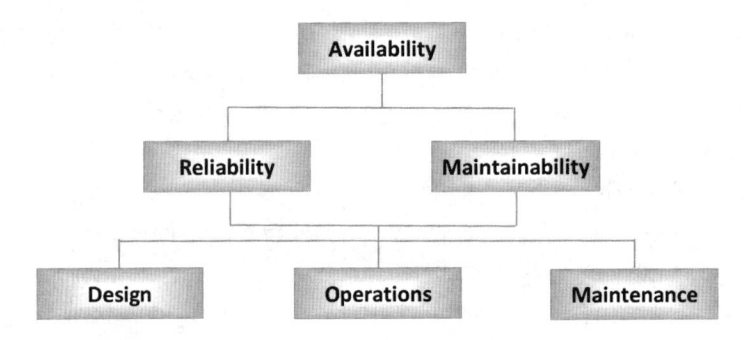

Fig. 7.3 RAM—Reliability, Availability, and Maintainability

Let us start with **reliability**. How much can we do about reliability? To answer this question, we need to go back to the three levels of reliability as shown in Fig7.2: physical asset reliability, system reliability, and process reliability.

As mentioned in Chapter 6, physical asset reliability is inherent as per the physical asset's design and is based on a specified operating context. Each physical asset comes with a designed mean time between failures (MTBF), which in turn derives a predicted failure rate (λ).

$$\lambda = 1/\text{MTBF}$$

The task of the maintenance function is to preserve that predicted failure rate under the specified operating conditions by performing the right inspection and maintenance tasks at the right frequency. The reliability of the physical asset is a function of the failure rate and is calculated as

$$R(t) = e^{-\lambda t}$$

where *t* is time.

We know that no physical asset operates on its own or under ideal (specified) conditions. Physical assets will be connected to other physical assets or components, which will have different reliability (or failure rates). Operating within those systems and in different conditions will produce other types of failures. Hence, we will have the system reliability, which will be based on the new failure rates. For example, an electrical motor will have a certain predicted failure rate (manufacturer's MTBF) under certain operating condition, and is going to be used on a conveyor belt. The motor is connected to a gear box, then to the drive pulley, and finally to a power supply source. The whole conveyor, as a system, will have a different reliability since the gearbox, pulley, and power supply will have different failure rates.

In operations, systems and subsystems will be put together to form a transformation process where there are inputs and outputs interacting. Each system and subsystem will have different reliabilities and on top of that, other inputs, such as human interface, raw materials variations, and many others, added to the interaction, will create other types of failure. Using the above example, the conveyor is now attached to a mixer and a packaging machine. Operators load the conveyor with raw materials to feed the mixer and the packaging machine. The whole process itself will have a different reliability.

System and process reliability can be improved by installing more reliable physical assets or asset systems in series, or by adding redundant physical assets/asset systems in parallel as shown in Fig. 7.4 below.

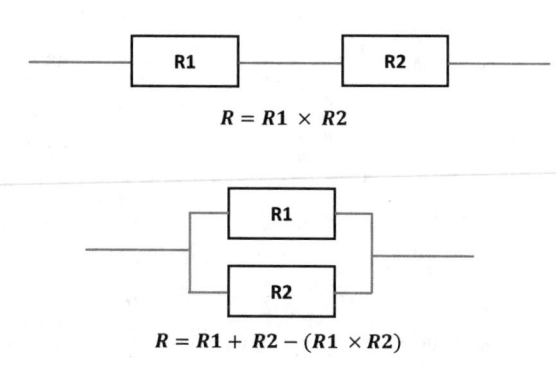

$$R = R1 \times R2$$

$$R = R1 + R2 - (R1 \times R2)$$

Fig. 7.4 Reliability of Systems in Series and Parallel

In the example of the conveyor belt it can be that you switch to a more reliable conveyor drive system, or you decide to install another conveyor belt in parallel in case the main conveyor belt fails.

Reliability engineering is tasked to ensure targeted reliability is achieved and maintained at all the three different levels.

Availability is the probability that a physical asset will operate satisfactorily at a given point in time when used in a certain operating context. In Chapter 5 we briefly mentioned availability as the ratio of runtime over scheduled time and how this ratio is used in the calculation of OEE.

From reliability engineering point of view, availability of a physical asset is calculated as

$$A = \frac{MTBF}{MTBF + MTTR}$$

where MTBF is the mean time between failure of the asset and MTTR is the mean time to repair. Availability must be calculated, tracked and improved at all three levels of availability to achieve the required operational availability:

- Inherent availability at the physical asset level
- System availability at the system level
- Process availability at the process level

How many times have we experienced repetitive failures of a physical asset which are not accounted for in a process downtime; just because the failed physical asset can be bypassed by other means - low physical asset availability versus high system/process availability, e.g. the conveyor fails and materials are transferred to the mixer via trucks to keep the process going.

How many times have we seen process downtime caused by failures of different physical assets – high individual physical asset availability versus low system/process availability, e.g. two hours process downtime last week was caused by 30 minutes conveyor jam, 15 minutes motor trip, 45 minutes gearbox failure and 30 minutes power outage.

The last item of RAM is *maintainability*. Maintainability is another design attribute of a physical asset and is the totality of design factors that allow maintenance to be accomplished easily, safely, and effectively. Maintainability is one of the most-neglected items in physical asset management, and is very often sacrificed at the expense of other attributes. This results in poorly designed physical assets with components that are difficult to access, dismantle, maintain, or repair. The impact is higher failure rate (lower MTBF) and higher MTTR, which affects both the reliability and the availability of physical assets.

With RAM in mind, physical asset management can help the maintenance management, reliability engineering, and operations functions achieve reliability, availability, and maintainability at the physical asset, system, and process levels. We know that the maintenance management function cannot improve the inherent reliability of a physical asset, but it can certainly sustain it to maintain consistent availability. We also know what reliability engineering needs in order to achieve certain specified system reliability, based on current operating context. And we know what level of availability & reliability operations needs from its transformation process. In all these areas, physical asset management can provide support ultimately to ensure that physical assets are designed/acquired with the right attributes at the outset to meet the required operational service level in a safe manner.

CHAPTER 8 —
Project Management

"No matter how good the team or how efficient the methodology, if we're not solving the right problem, the project fails."
~ Woody Williams

The project management function is heavily involved in four of the six physical asset lifecycle phases: planning, creation/acquisition, installation, and commissioning. This may sound weird to some, but believe it or not, project management is, in my opinion, one of the most important functions in a physical asset lifecycle. There are a few key reasons for this, because project management is:

1. when everything starts for a physical asset
2. when great things get created and built, but is also the stage when many other things can go wrong
3. when, if the wrongs are not corrected, the physical asset owners can get stuck with those wrongs for years
4. having a significant impact on the total lifecycle cost of the physical asset

The amazing part of all this is that the project management function is the least spoken about when we discuss physical asset management. There is a huge dissociation between the two functions in many organizations.

"Why is that so?" you wonder.

Well, first of all, at the project management stage the physical asset is not yet realized and, at that point, not much information is known regarding what the physical asset is going to look like. The concept of the physical asset or asset system slowly evolves into a collection of ideas, and finally evolves into the actual physical asset. During the evolution and creation of physical assets, many changes can happen. These include things that can have a major impact on the different phases of the physical asset lifecycle, such as performance standards, reliability, maintainability, and even lifecycle costs.

Secondly, the focus of the project management function is mainly on its ToF—such as cost, time and scope (project objectives)—and not much on the value to the organization (project benefits). All projects have objectives and benefits. While both are very important to project management, project objectives should not take over project benefits. I should quickly point out that project benefits can be hugely influenced by the organizational structure in which the project is executed. A projectized structure will focus mainly on project objectives, while a functional structure will fancy project benefits. A matrix structure with a blend of functional and projectized characteristics will provide a good balance for both and would be ideal. Regardless of the structure what is the point of successfully executing a project on time, within scope and budget, when the physical asset/asset systems installed encounters repetitive failures and design shortfall soon after? Historically, the focus has been on objectives that represent the outputs of the project and are mostly short term achievements. Project benefits, on the other hand, represent the outcomes of the project—the ones identified in the needs assessment or project charter. They are the reason(s) the project is being carried out, and they represent the long-term objectives. These long-term objectives must align with the strategic goals of the organization and must make a meaningful impact.

The encouraging thing is that organizations are slowly coming to realize the impact of good project management in the acquisition, installation and commissioning lifecycle phases. As a result, we get to see physical asset management philosophies and strategies being incorporated in the initial stages of the project management process, via engaging stakeholders and cross-functional teams early on in the project.

Let us take a look at the project management process, itself. We all recognize that the execution of capital projects is a very key function in any organization. The function is mainly achieved through the application of project management methodology and industry best practices.

As per the Project Management Body of Knowledge (PMBoK) from the Project Management Institute (PMI), *project management* is defined as "the application of knowledge, skills, tools and techniques to project activities to meet project requirements."

Requirements (project objectives) are usually met when projects are delivered on time, within budget, and within agreed upon scope. This can be achieved by applying multiple project management processes over the lifecycle of the project. These processes fall under five different groups:

1. The *initiation process group* entails processes to define and start new projects
2. The *planning process group* is where the project scope is defined and work activities are identified
3. The *execution process group* includes processes required to complete the project
4. The *monitoring and controlling process group* defines the processes needed to track, review, and monitor progress of the project
5. The *closing process group* is the processes that are in place to finalize and formally close the project.

The process from each process group is linked to the process from other process group, via outputs that then become their inputs, as shown in Fig. 8.1. Through interactions between the different process groups, there is seamless transition, many back and forth, and sometimes overlaps, over the whole lifecycle of a project.

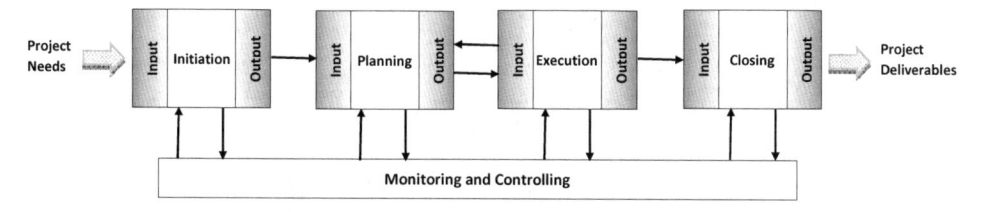

Fig. 8.1 The Project Management Process Flow

The purpose of capital project management in organizations is to upgrade, replace, add, or dispose of physical assets to meet certain operational needs. Operational needs can be of different types, such as:
- Capacity expansion, reduction, or consolidation
- Renewal or updating of existing capacity
- New business direction
- Process improvement
- Technical developments
- Environmental or safety improvements
- Response to regulatory changes

Capital projects involving physical assets can also be of different sizes, varying from a simple physical asset refurbishment, to a like-for-like physical asset replacement, to the addition of a completely new process, depending on the types of needs listed above.

Irrespective of the type and size, all projects carry a certain element of cost and risk, which varies over the capital project lifecycle. Generally, cost is low at the start, ramping up as the work progresses, and dropping significantly as the project gets completed. Risk and uncertainty, on the other hand, are at their highest at the start of a project, decreasing gradually as decisions are made and accepted, and as the project progresses. So it is very important to have a good balance between cost and risk over the whole project lifecycle. The right decisions must be made at the right times, especially at the planning, acquisition, and installation and commissioning stages of a project lifecycle, when it is less costly to make any changes. As mentioned above, organizations are becoming aware of this caveat and the right stakeholders are identified early on in the project, to get involved and have a say about physical asset design and installation. Furthermore, physical asset management philosophies and strategies—such as reliability-centered design, system engineering, value engineering, and configuration management—are applied early on in the project.

Envisage now in Fig. 8.2 below how the lifecycle phases of a physical asset look like when mapped onto a project's typical lifecycle phases, which also have overlaps. As you can see, it can become quite a nightmare to draw any line between the physical asset management function and the project management function.

For example:
- Where do we draw the line between the physical asset management planning phase and the project management initiation phase?
- How much of the feasibility studies, scope decisions and preliminary design should be completed before handing it over to project management to execute?
- In what shape and form should the handing over happen?

These are just a few of the many questions that can arise when it comes to transitioning from one phase to the other, or from one function to another. Important details can be missed or get altered, which can have a significant impact on the physical asset or project lifecycle and, ultimately, on the organization.

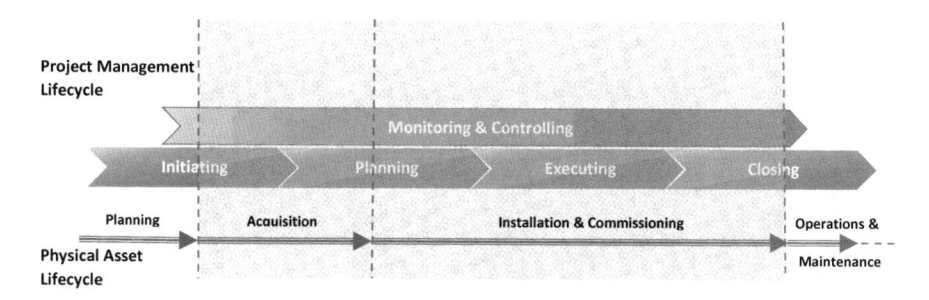

Fig 8.2 Project Management and Physical Asset Lifecycles

From Fig. 8.2, we see that the project management function is involved in at least three phases of a physical asset's lifecycle, namely planning, acquisition, and installation & commissioning. It is also understood that in the acquisition and installation & commissioning phases the project management function are fully involved and even assume the lead in delivering physical assets or asset systems.

Coincidentally, these are the two most critical phases of a physical asset lifecycle in terms of cost and risk. The acquisition phase is where all the design details are finalized, and crucial decisions are made. As mentioned before, most of the risk and uncertainties lie in that phase. The installation & commissioning phase is where the needs and designs are actually transformed into actual physical assets/asset systems. These two phases are very important and are literally the "make or break" stages of the development of any physical asset/system, from cost, function, reliability, and risk perspectives. As shown in Fig. 8.3, the vast majority of the total lifecycle costs of capital projects may already be permanently incurred prior to the commissioning stage, meaning acquisition and installation have already happened.

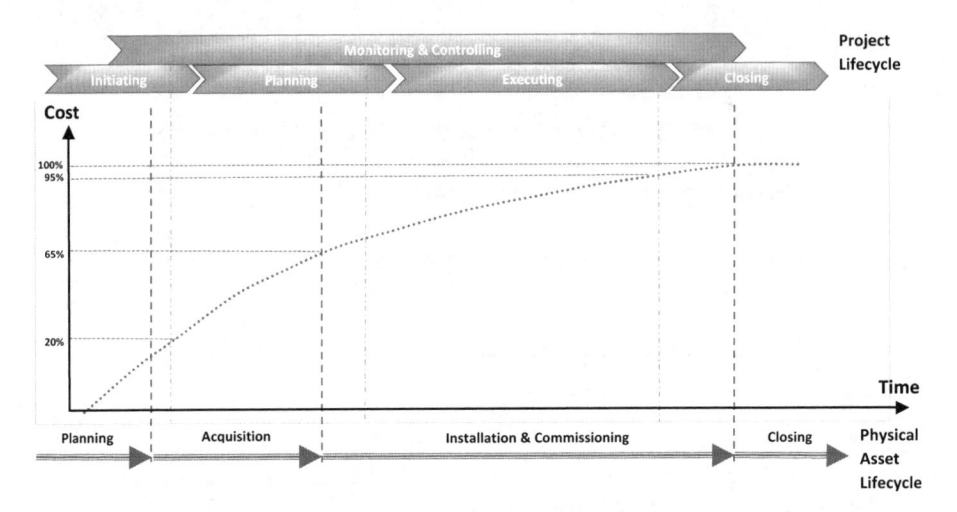

Fig 8.3 Project Lifecycle, Physical Asset Lifecycle and Cost

Historically, design has often been classified and restricted to the technical attributes of the physical assets, rather than a more holistic approach to the functions and risks associated with the design. Due to the fact that design is the first major cost-related element at the project management level, it is obvious to say that a good economical, optimum design is critical for an effective holistic physical asset management start.

Going back to Figure 4.8, it can be seen that the project management function needs to work closely and interact with different functions to achieve common organizational objectives:
- working with operations to ensure operational needs are included in project scope, and that timelines are met when delivering projects
- working with maintenance management and reliability engineering to ensure RAM principles are included at the designing stage, and that proper handing over of physical assets is carried out during the closing phase

- working with finance to ensure funding is available, projected cash flow is adhered to, and projects are completed within allocated budget

If you look at organizations in the recent years, project management and execution of capital projects have been facing huge challenges. You may argue that this is not the case, based on all the projects that have been or are being completed with the application of project management best practices. But, let me tell you, if you look closely at those projects, you will notice gaps and issues that impact physical assets performance, service levels to customers/communities at large, and ultimately the whole organization. As mentioned in the MGI's 2013 report *Infrastructure productivity: How to save $1 trillion a year*, organizations need to improve on project selection and delivery, and need to make the most of existing physical assets. In other words, ensuring both project objectives and project benefits are achieved in the most productive manner.

"How can such things actually happen in reality?" you ask.

Well, there are many such project fiasco examples that we can learn from, if we do some research. Besides the common project delays and budget overruns, projects are often completed with either overdesign or underdesign issues, or deficiencies resulting in repetitive failures with potential warranty/legal issues.

Ultimately, all this will lead to either inability to provide the required service level, or high cost to rectify deficiencies, or high costs to operate and maintain. To understand the reasons behind those situations, let us first review the three groups that normally handle and execute capital projects in organizations:

- *Maintenance management* — in many organizations, the maintenance management function handles small- to medium-sized capital projects that are recurring or one-time type of projects.
- *Project management/engineering* — their main function is to design and execute capital projects of different sizes (mostly medium & large) and types

- **External agencies/outside contractors** — execute medium to large projects of all types.

In other words, the three groups make changes to physical assets and asset systems in organizations. All these parties do a great job of executing capital projects and fulfilling all the project management requirements.

The question at hand is, do they actually fulfill the business requirements?

If you <u>agree</u> to at least one of the eight statements listed below, you have answered **NO** to the question at hand.

1. have not been involved in process
2. Completed projects are dropped on the lap of operations and maintenance by the capital project group, without proper handing over.
3. The project is delayed and operations have to find alternate ways to meet customers' demands to make up lost production time.
4. There is no or missing documentation such as equipment manuals, engineering drawings, etc. during the commissioning and closing phases.
5. Poor design and selection of physical assets result in poor reliability and maintainability.
6. There is lack of consideration for standardization of physical assets and spare parts
7. There is oversight of total lifecycle cost; i.e., cost to operate, maintain, upgrade or energy consumption.

Case Study 8.1

The Maribonsa Pasta plant has been undergoing a significant expansion over the last year or so. A new production line with new equipment has been added to process a new product that the marketing department has been pushing for and which will bolster sales for the company. Operations staff and the engineering group, which takes care of all the capital projects, have been working hard for the last few months to put the finishing touch to the installation and get the project completed on time to launch the new product. Maintenance has also been involved as and when required during the installation & commissioning phase, and has helped out to shutdown equipment, complete some minor tasks, and support the project overall.

Commissioning of the new installation has been a rough ride, especially due to the fact that they have struggled to meet the product specifications and achieve the required throughput. Finally, after several test runs and innumerable system tweaks, the commissioning was completed and the project was closed. Launching of the new product was on schedule and operations initially started at 75% capacity to slowly ramp up to full capacity over the week. Everybody was happy with the new product launch and the plant was running at full steam.

As in any type of operations, production downtime happens and equipment does fail. Maintenance has been called in every time to fix the equipment failures and make minor adjustments. After a few weeks of the same problems occurring, operations was not happy with the performance of maintenance and their inability to resolve the issues. Maintenance, on the other hand, has been complaining to operations that they do not have the required expertise on the new equipment and they do not have the required documentation to do a better job. The issue has been escalated to senior management to resolve because it has started to impact production and external customers.

What they found out was that, at the time the project was closed, a few key items were not addressed:
- Not all equipment manuals were delivered
- The equipment supplier did not supply a maintenance program at project close out, as per contract
- No formal training of maintenance staff took place during the close out phase
- The new equipment has not been entered in the CMMS system

- There is no maintenance program in the CMMS for the new equipment
- There is a lack of technical support from the equipment supplier.

The above depicts a real-life scenario that many of us may have experienced, either as operations staff, as maintenance staff, or as project managers executing capital projects. I can assure you that business leaders and managers would not like to be in such situation, where organizations get stuck with expensive physical assets for years, without the know-how and documentation to operate/maintain them to realize maximum value. The case study is not meant to point fingers at anyone, but is included here to highlight major organizational gaps when it comes to executing capital projects and delivering physical assets.

Physical asset management can support in the process of capital project execution to ensure that the project management function bridges the gaps existing in the lifecycle of physical assets, such as:
- The final design incorporates other functions' needs and requirements
- Scope changes during execution are shared with other functional areas for approval
- The contract is drafted with physical asset technical specs and performance standards
- There is a clear definition of contract close-out with regards to warranty
- A complete physical asset list is provided, along with related documentation
- A maintenance plan is developed, outlining maintenance tactics and strategies, together with list of critical spare parts.
- Physical asset reliability data and statistics are provided
- Commissioning and training of personnel from operations and maintenance takes place

Let us now take a look at how the physical asset management and project management functions can work together on the three areas of focus to ensure, both project objectives and project benefits are achieved, while at the same time ensuring the holistic physical asset management philosophy is preserved.

Scope — project scope management includes the processes required to ensure that the project covers all the work required, and only the work required. It consists of:
- Collecting and documenting all needs and requirements
- Defining clearly the scope of work
- Creating work breakdown structure to identify distinct deliverables
- Verifying and accepting deliverables as per set requirements
- Controlling the scope of work to ensure no scope creep

In these processes, there must be lots of interaction and involvement between the two functions for successful project scope management. Any scope change that affects the design, the function, or any other attribute that can impact the service level or operational needs, has to be brought to the attention of the physical asset management function. The impact can be huge on all aspects of the business case: lifecycle cost, RAM, risks, and performance standards.

Time — project time management includes the processes needed to manage timely completion of the project and consist of:
- Defining the activities and sequencing them
- Estimating activity resources and durations
- Developing a schedule, taking into consideration all constraints
- Controlling the schedule to monitor progress and manage any delays.

Delays in project delivery not only have an impact on project cost, but may also create a loss of business to the organization. They may result in loss of production, loss of sales, customer complaints, and many more. From a strictly physical asset perspective, as projects are delayed, the conditions of physical assets continue to deteriorate, leading to unexpected failures and creating reactive situations. Again, both physical asset management and project management can benefit greatly from working closely together.

Cost — project cost management includes the processes that will help to complete the project within budget, and consists of:
- Estimating costs to complete project activities;
- Determining budget, based on activities cost estimates;
- Controlling costs to monitor spending and manage variances

Cost and budget estimates, if not accurate enough, can have a huge impact on funding strategies and cash flow. Organizations do not have an infinite amount of funds to inject into projects. Accurate forecasting and planning are required. Cost, if not controlled, can also have a negative impact on the business case and the rate of return of the project. And any cost overrun will increase the lifecycle cost of the physical asset. Physical asset management has to work with project management to ensure optimum cost is incurred, with accurate forecasting, without compromising any other features of the project/physical asset.

CHAPTER 9 —
Procurement Management

"Don't tell me where your priorities are. Show me where you spend your money and I'll tell you what they are."
~ James W. Frick

Procurement is another very important functional area of any organization. It has quite some influence on the physical asset management function. **Procurement** is the process of acquiring goods & services from an external source. The goods and services can be raw materials, labour, physical assets, components, or consulting services. When we consider physical assets, procurement refers to the processes involved in engaging outside (and possibly inside) sources to:
- Execute work (construction, fabrication, etc.)
- Provide services (consulting, management, engineering, etc.)
- Supply materials and products
- Design, manufacture, and/or supply equipment

The whole procurement process can range from supply of simple components worth a few hundred dollars to installation of multi-million-dollar system of physical assets. All physical assets of an organization go through some type of procurement or acquisition process. The main purpose of the process is to acquire those physical assets at the best possible price, as per the specifications of the

requester, after consideration of the alternatives' ownership, risks, and benefits. The procurement process is usually well defined so that needs and requirements are clearly understood and fulfilled under a certain level of control to ensure:
- Organizations get value for money in the transaction
- All applicable policies and by-laws are complied with
- All risks and terms & conditions are well written and spelled out
- Exposure to fraud and collusion is minimized

While some procurement process may be simple as shown in Fig. 9.1 below, others can be quite cumbersome and lengthy depending, on the magnitude and complexity of the acquisition. Things such as legal matters, fair monitoring, multi-stage procurement, bid committees, and many more can come into play, which may slow down the process. But, as listed above, there are many benefits derived when following the procurement process, even when facing time and resource constraints.

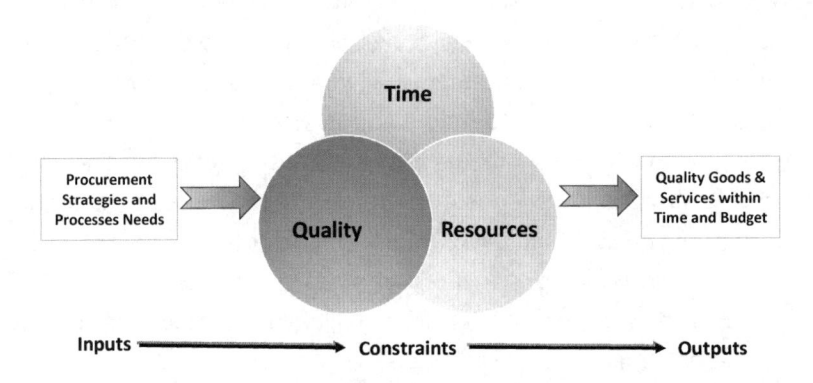

Fig. 9.1 The Procurement Process

From an input perspective, all needs are clearly defined and accurately estimated, all roles and responsibilities are clarified, all outputs are clearly identified and the reporting requirements agreed upon. The benefits generated from the procurement process are:
- Low cost

- Right quality
- Right specs
- Unbiased vendor selection
- Best value for money

Achieving all those benefits can be a tricky task in the sense that the procurement process will be different, based on the type of organization, the policies in place, the purchasing strategy adopted, and the nature of the acquisition.

In the private sector, the procurement process can be straightforward, with obtaining a few quotes, reviewing, and selecting the right fit for the task, based on certain evaluation criteria. In the public sector, on the other hand, there is more sensitivity in how the process is handled; stringent procurement policies drive the whole process. Most procurement activities go through a lengthy tendering process, with strict evaluation and selection criteria to award the winning bid. In both the private and public sector procurement processes there must be close interaction and effective communication between the purchasing function and the client/requesting department throughout the whole purchasing cycle. Fig. 9.2 below depicts the purchasing cycle for a physical asset.

Need
- Identify requirements
- Develop estimates
- Secure funding
- Get approval

Purchase
- Analyze options
- Select procurement strategy

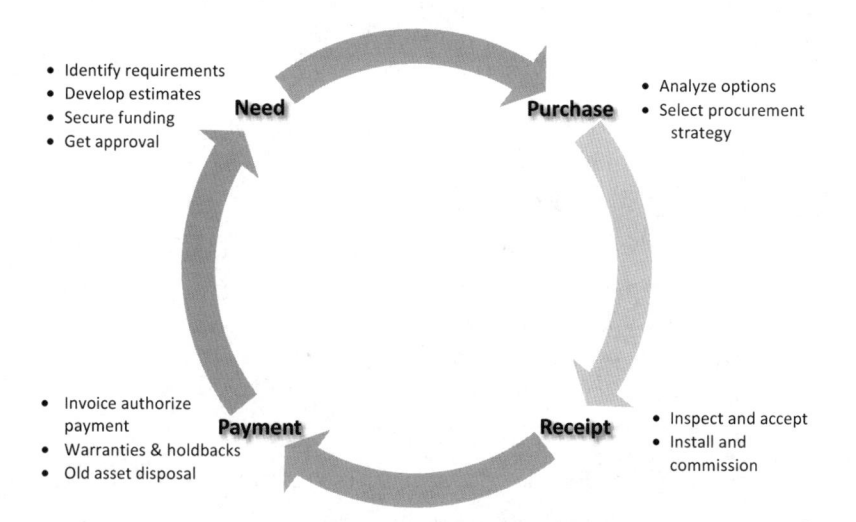

Receipt
- Inspect and accept
- Install and commission

Payment
- Invoice authorize payment
- Warranties & holdbacks
- Old asset disposal

Fig. 9.2 Physical Asset Purchasing Cycle

At this point, it is good to highlight the difference between procurement and acquisition. So far we have been talking about the procurement process, which is the act of buying goods and services by following a purchasing cycle based on specific client's needs. Physical assets can be procured through the purchasing cycle if they are off-the-shelf items. However, when they are not off-the-shelf physical assets or asset systems, then the procurement process becomes more prominent in the acquisition phase of a physical asset lifecycle. *Acquisition* is the act of conceptualization, initiation, design, and development of physical assets or asset systems. Acquisition is a wider concept than procurement, and it covers a bigger scope in physical assets or asset systems lifecycle.

The acquisition process can typically involve the initial phase of needs identification, design, development, testing, etc. This is demonstrated in Fig. 4.8, where procurement is heavily involved in the acquisition phase of physical asset lifecycle. The key aspect of procurement in the acquisition phase is the procurement strategy. The procurement strategy is very important to acquire the right physical assets or physical asset systems via a competitive process. There are numerous forms of procurement strategies and there is no one best strategy that fits all. Some factors that influence the procurement strategies are:

- Owner requirements
- Market conditions
- Time, cost and quality requirements
- Quantity/size of work
- Type of work
- Resource availability
- Technical complexity
- Risks and allocation of risks
- Budget
- Availability and abilities of outside sources

Many organizations are faced with the dilemma of whether to do the work, or part of the work, in-house or to contract out, or a combination of both. Both options have advantages and disadvantages, and the strategy will ultimately depend on many of the factors listed above. Some of the common outsourcing procurement strategies are:

- Traditional construction
- Detailed design and construct
- Design-develop-construct
- Design-construct-maintain
- Build-own-operate-transfer
- Project management
- Construction management

All the above have pros and cons, and will be dependent on the needs and requirements of the client, the amount of control they want, the costs involved, the flexibility to make changes, and the availability of resources. Any particular strategy should be tailored to the client's and project's needs. Below are some guidelines that can help organizations select which procurement or contract strategy they would adopt to suit their needs.

- **Traditional construction** is where the client is prepared to take control of the project, using internal staff or using consultants. The client then hires the necessary contractors and subcontractors for the work.
- In a **detailed design and construct strategy**, there is less involvement of the client. The client develops the concept design and the rest is taken care of by the contract.
- The **design-develop-construct** contract is where the client just provides a brief detailing of the objectives to be met, and then a general contractor is hired to take care of the concept design, detailed design, and construction.
- In **design-construct-maintain**, the contractor is also required to take care of the physical assets through proactive maintenance to keep them in a state of good repair.
- In **build-own-operate-transfer**, a private entity receives a concession from a public or private organization to finance,

design, construct, and operate a facility. At the end of the term of the contract, the physical assets are transferred to the client.
- In the ***project management*** context, the client engages an agent (project manager) to assist with carrying out the project and achieving the client's set objectives.
- In the ***construction management*** strategy, the client hires a construction manager (a consultant or contractor) to provide services for the construction phase.

Irrespective of the procurement or contract strategy adopted, a competitive or non-competitive procurement process has to be followed, as per a purchasing cycle, with the involvement of internal and external resources interacting with other functional areas.

"But who uses these procurement strategies in organizations?" you ask. "Who decides which strategy to apply and when?"

Well, the nature of procurement for physical assets or asset systems, initiated by the different functional areas in an organization, will dictate the procurement process and strategy to adopt. For example, the maintenance management function will be purchasing smaller-in-size physical assets or parts/components of physical assets, and may put a tender out for potential suppliers to bid. Operations who will require more office space may hire consultant to work on a detailed design before proceeding with construction. Or the project management function in charge of managing the plant expansion project may take a design and construct approach. For those of you who enjoy graphical illustration, Fig. 9.3 below shows a simple competitive procurement process.

Define need and scope specifications → Develop, review and post call document → Review and evaluate bid document → Award contract

Fig. 9.3 Competitive Procurement Process

For the most part, a competitive procurement process will follow the following steps:
1. Defining the needs

2. Preparing the specifications
3. Getting approval
4. Preparing the call document
5. Bidding period
6. Reviewing and evaluating bids
7. Awarding contract

Case Study 9.1

Raskoltar Inc. is a transportation company that has a fleet of 50 trucks, which provide same-day delivery service to a number of customers in its neighborhood. The lifecycle of these trucks is six years and, as per the company's policy, all trucks need to be replaced once they reach six years of age. If the trucks are not replaced by then, Raskoltar Inc. may incur more unplanned trucks downtime, which will affect their service level. At the same time, maintenance and repair costs will eventually increase.

Raskoltar Inc. has identified ten trucks that will reach six years in December of this year, and has started the procurement process already since December of last year. Business cases have been developed and signed off. Funds have been identified and approved.

Typically, the turnaround time from submitting business cases to the procurement division to the final delivery of the trucks is between 12 and 15 months. If everything goes well, the trucks will be delivered on time and meet the six-year lifecycle timeline. However, in the 12 to 15-month period, lots of things could happen that could extend the time frame by months.

In this case, there were a few glitches with the bid, which resulted in the call document closing deadline being extended twice, adding a further 2 months to the process. Furthermore, once the contract was awarded, there was delay in delivering a portion of the ten trucks due to manufacturing issues.

The result is that Raskoltar Inc. has had fewer and less reliable trucks on the road to provide service. On top of that, the company has had to incur higher maintenance costs and put up with an increasing number of customer complaints due to lower service level. Can Raskoltar Inc. afford the risk of losing customers due to delays in the procurement process? Should they contract out temporarily part of their business to meet customer demands and incur more losses in the process?

There are many potential risks during a procurement/acquisition process with lots of things that can go wrong, which can delay the process, affect the scope and quality of the deliverables, or impact the budgeted cash flow. No organization wants to end up with poorly designed physical assets, or asset systems with lots of deficiencies and performance issues, which can result in service level impact, shorter than expected economic life, higher maintenance costs, and higher lifecycle costs. No organization would want to experience huge delays in the procurement process that can cost significant amounts of money in terms of penalties, wastes, legal fees, or loss of business.

During the procurement/acquisition process, it is very important to have all specifications, requirements, and performance standards well documented. As illustrated in Figure 4.8, the procurement function is involved heavily in the acquisition phase of a physical asset lifecycle. It is very important that all the functional areas concerned with the acquisition work together to ensure all requirements and needs are documented and stipulated well upfront, to avoid complications and surprises in the procurement/acquisition process.

Of course, this is easier said than done. Very often, we see that those specifications, requirements, and performance standards are either not clearly documented, or keep chnaging, or are outright missing. How many of us specify what MTBF we need when we purchase a piece of equipment? How many of us stick to standard physical asset types so as to leverage on service and spare parts? How many of us look at lifecycle costing when we design an asset system—how much will it cost us to operate and maintain over its whole lifecycle? How many of us include specifications around energy conservation?

We do realize and understand that there are many constraints around the procurement/acquisition process that organizations have to put up with, such as complex purchasing policies/by-laws, low bid policy, preferred vendor policy, sole sourcing requirements, or unavailability of specified physical asset. That being said, the physical asset management function can work with the other functional areas to help:

- Ensure the right specification and performance standards are included in the procurement documentation.
- Enforce opportunities for configuration management, standardization and value engineering.
- Provide better physical asset planning, with enough lead time for the procurement/acquisition process to happen.

CHAPTER 10 —
Financial Planning

"Think ahead. Don't let day-to-day operations drive out planning."
~ Donald Rumsfeld

Organizations, both in the public and the private sector, need capital funds to invest and reinvest in their physical assets. Capital funds, in general, are scarce in organizations, especially in the current fiscal situation. Therefore, organizations need to know where and when to invest the money and need to prioritize their needs over wants. Capital investment in physical assets has to be planned carefully so that funds are made available at the right time, in the right amount, and for the right reasons. It is not a straightforward exercise, and this is where good financial strategy and planning comes into play.

First, we need to understand the sources of funding available to organizations, the cash transactions, and the cash flows.

Fig. 10.1 Cash Flows for an Organization

The organization in Fig. 10.1 will rely upon capital investors to provide necessary capital funds to invest in physical assets and infrastructure. To make the right decisions on what funding sources to use, managers and senior leaders in organizations discuss and debate questions such as:

- Does operations generate sufficient funds to enable the organization to use for reinvestment or will they need to borrow?
- Where will the money for physical asset acquisition come from? From operations or from borrowing?
- Is the organization solvent? That is, has it sufficient liquid assets to meet its short-term obligations?
- Why does the organization need to borrow so much when it's making so much profit?

"Fair enough," you say. "So what are the actual sources of capital funds for the organization?"

Well, in the private sector, those capital funds can come from the shareholders, profits/cash, loans, or new investors. In the public sector, the capital funds can be generated from property taxation, revenues, other government-level grants and subsidies, capital reserves, or debts. When organizations plan for funding and decide what funding sources to use, there a few key factors that are considered, such as internal rate of return, net present value, interest rates, debt ceiling, or taxation rate, which will impact the organization's cash flow. Lots of work is accomplished behind the scenes for the right financing strategies to come up with a balanced budget and a sustainable cash flow.

Planning

Once the organization has decided on its financing strategies and identified the potential sources of capital funds, they have to start working on planning where it's needed, why it's needed and when it's needed. The process consists of three distinct phases:
1. Asset planning phase
2. Capital planning phase
3. Financial planning phase

Within those three phases, there are many discussions around what projects or investments get the go ahead and which ones are going to be deferred for later years. The key reason for this is because capital funds are limited compared to the number of investment needs and competing priorities. To get a fair share of the pie in the form of capital funds within the requested time frame, answers to these questions are required: What? Why? When? How? The whole process is illustrated in Fig. 10.2 below.

Physical Asset Management Planning Process

Asset Planning — WHAT? WHY?
- Identify needs based on demands of stakeholders
- Identify needs based on conditions of assets
- Assess and optimize needs
- Create list of projects and prioritize
[List of capital projects]

Capital Planning — WHAT? WHEN?
- Complete feasibility studies and business cases
- Decide on best options and when needed
- Obtain cost estimates
- Plan project based on priorities
- Develop capital plan
[Multi-year capital plan]

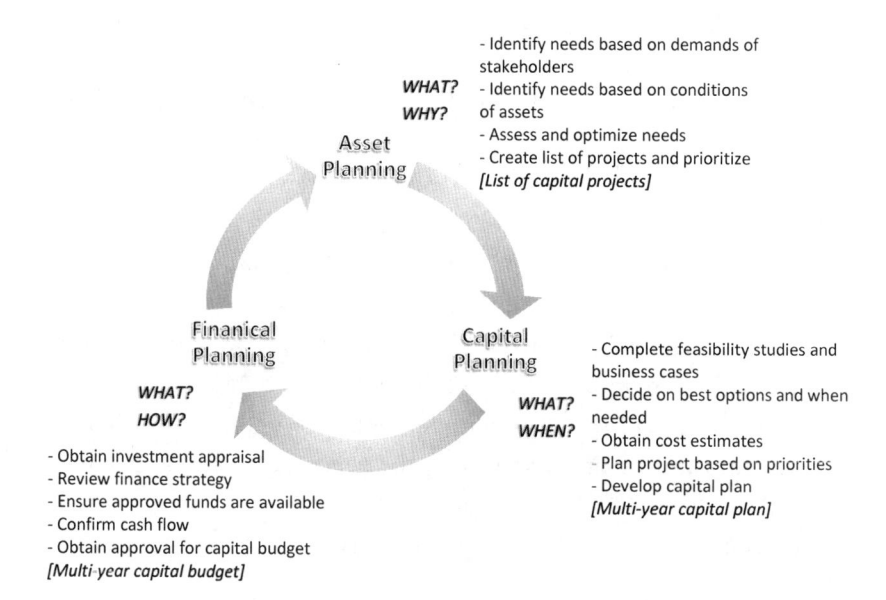

Finanical Planning — WHAT? HOW?
- Obtain investment appraisal
- Review finance strategy
- Ensure approved funds are available
- Confirm cash flow
- Obtain approval for capital budget
[Multi-year capital budget]

Fig. 10.2 Physical Asset Management Planning Process

The *asset planning* phase consists of identifying needs based on current condition of physical assets, their remaining useful life, and the short-term and long-term needs of the organization.

Organizations rely on capital investment in physical assets to expand, to enter a new line of business, to replace due to technological obsolescence, or to simply keep them in a state of good repair (SOGR). To be able to prioritize and allocate funds to the right needs at the right time, organizations need to know the conditions of their physical assets, their remaining life, their operational requirements both for short-term and long-term, and must make rational capital investment decisions. Investment decisions can be grouped in six general categories:

1. Purchase of new physical asset or facilities, which is generally tied with growth.
2. Replacement of existing physical asset or facilities, which can be prompted by age, condition, lifecycle cost, or obsolescence.

3. Make or buy decisions revolving around core competency of the organization or amount of control required.
4. Lease or buy decisions, which depend on the initial capital outlay and short- and/or long-term needs.
5. Temporary shutdown or plant-closing down decisions, which depend on the operational strategy.
6. Addition of new product or service decisions, which are tied to the competitive priorities of the organization.

Once those decisions are made and prioritized, the organization knows the fate of its physical assets in the short- and long- term. This is referred to as *physical asset planning* and is a cornerstone in the financial planning process to fulfill the needs of capital investment in a timely manner. Organizations need to have a proactive physical asset planning process to be able to forecast needs for capital funds.

The next phase is *capital planning*. Capital investment needs—such as short-term and long-term renewal/rehabilitation tasks, replacement decisions, disposal requirements or any new additions—are translated into capital projects, based on the conditions of existing physical assets, or based on project needs of the organization. These capital projects are backed up with solid business cases, investment appraisals, feasibility studies, and cost-benefit analysis, to come up with budgetary estimates. Budgeting encourages planning. Without knowledge of where the organization is heading with its physical assets, it is hard to plan for resources. The "wish list" of capital projects need to be further prioritized with management reviews to make final decisions. Options and alternative scenarios are discussed, based on the impact on service level and risks to the organization. The capital investment projects are then integrated into a capital plan, with budgetary estimates, which then translate into a multi-year capital plan with at least a ten-year planning horizon. Traditionally, organizations have their capital forecast spread over many years ranging from ten to thirty years, and sometimes extending to the lifespan of the physical asset based on condition assessments and future needs. Fig. 10.3 below shows an example of a multi-year capital investment forecast.

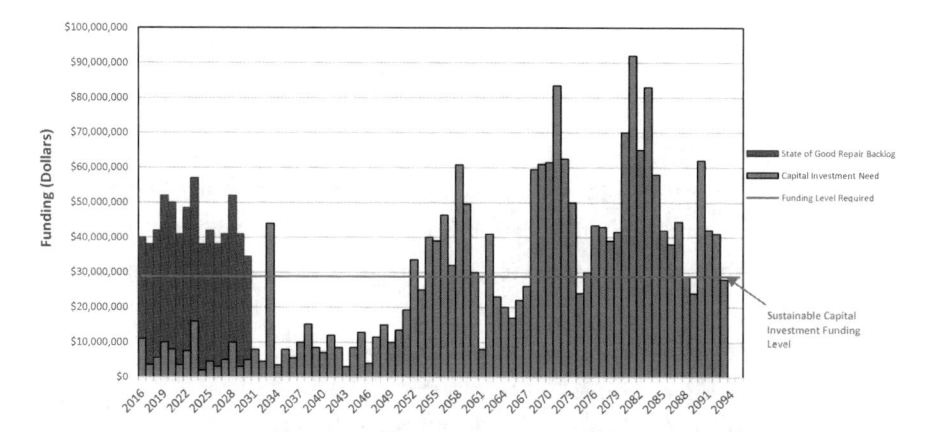

Fig. 10.3 Multi-Year Capital Investment Forecast

The darker portion of the graph on the left the state of good repair backlog, which are typically unfunded physical asset needs. The horizontal line that runs across the graph depicts the required funding level in order to sustain a state of good repair over the multi-year timeframe.

The next phase, after the capital budget is presented is for the finance function to complete their planning. The main purpose of *financial planning* as a key function in an organization is to ensure that capital funds are available every year to cater to the identified and planned capital projects as per approved budget. The financial plan will take into account any revenue and funding sources and will identify any funding shortfalls relative to financial requirements. Cash flow is a key component of financial planning and is the essential pulse of any organization. The cash flow statement details various inflows and outflows of money through the organization. Cash flow can come from operating activities, from investments, or from financing. In order to ensure enough capital funds are available to execute the capital plan as per the budgeted cash flow, organizations (finance) need to manage closely the inflows and outflows from the different sources and develop a finance strategy.

"Hold on a minute," you say. "That's what all organizations do every year, don't they?" you ask.

It is true that all physical-asset-intensive organizations go through the capital budgeting process every year, but unfortunately the capital plan is not always successful. What we experience in many organizations these days is an imbalance between needs and wishes. On one side, needs are not clearly and fully identified, and on the other side, we tend to focus more on wishes than on real needs. Both combined lead to:

- Fluctuating capital investment requirements
- Inclusion of poor estimates in the capital budget
- Wrong projects eating up the bigger share of the pie
- Last minute rush for more capital funds

The result is injection of capital funds in the wrong physical assets for the wrong reasons.

In the private sector, financial planning and capital investment are focused mainly to increase revenue and to drive profitability. The result is high investment in physical assets to increase throughput, to launch new products, or to embrace new technology. If you recall in Chapter 2 we talked about equipment-centric organizations and in Chapter 4 about high-intensity maintenance physical assets being mostly in the private sector. As a result there is faster turnaround with regards to rehabilitation or replacement of physical assets (sometimes physical assets are replaced way before they reach the end of their economic lives) to meet organizational strategic objectives, or changing business landscapes.

In the public sector, the situation is different with infrastructure-centric organizations and low-intensity maintenance physical assets of high replacement value. In many countries from 1950 to 1975, the focus was mainly on building and adding new physical assets/infrastructure to match population growth and the rate of urbanization. This was a period of intense capital investment. Thereafter, there was a sharp decline in capital investment up until the year 2000. Clearly, all levels of government were under-investing in the maintenance and rehabilitation of their capital stock. With the now ageing and rapidly deteriorating physical assets, the public sector has been trying to catch up with capital investment. Currently, all levels of governments in many countries are facing a tremendous fiscal challenge in the form of investment deficit to cope with the ever-growing backlogs of maintenance, renewal, and replacement of ageing capital stock. This is causing financial stress on organizations and jeopardizing the sustainability and affordability of services. For example, in Canada and United States, we are witnessing a growing deficit in capital investment as shown in Fig. 10.4a and Fig. 10.4b.

Where do we find the money to fund the physical asset/infrastructure gap? How fast can the gap be filled? What are the risks looming on us, on our communities?

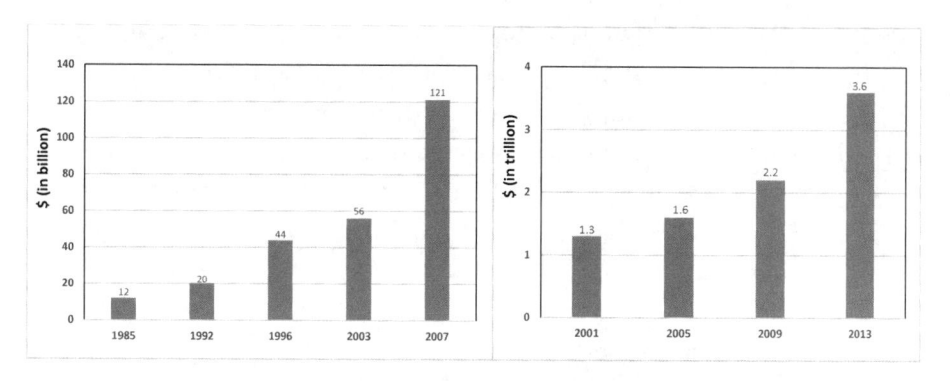

Fig. 10.4a / Fig. 10.4b (Source: various)
Capital Investment Deficits

In light of the above, it is very important that all organizations have a good financial strategy in place to source the required funds to inject in their physical assets capital plan and address the investment gap. One key step in this process is to have access to financial information about the capital stock, and the utilization and condition of the physical assets.

Financial Reporting and Control

Every organization, be it in the private or public sector, has to report on their finances in the form of a financial statement. Financial statements include financial information for external users of the information such as investors, regulators, and suppliers. Financial statements are usually audited and are prepared in accordance with generally accepted accounting principles (GAAP). GAAP are set by various standard-setting organizations such as Accounting Standards Boards and other accounting regulatory bodies so as to comply with the International Financial Reporting Standards (IFRS). One of the key items in a financial statement is the tangible assets of an organization. Financial statements should report on how physical assets contributions are managed and their impact on the organization's bottom line.

For example, valuation of public physical assets/infrastructure is regulated by the Public Sector Accounting Board, PSAB 3150, in Canada, Governmental Accounting Standards Board, GASB 34, in the United States, and similar standards in other countries. PSAB 3150 requires that all physical tangible assets to be reported at historical costs over their estimated useful lives and should include:

- Beginning-of-year and end-of-year balances
- Capital acquisitions
- Sales or other dispositions
- Depreciation expenses

Fig. 10.5 below maps the requirements to achieve financial reporting for physical assets in organizations. In the public sector, physical assets reporting should include an inventory of physical assets, the valuation of each asset, how they are amortized over their economic life, residual investment, and how changes to the physical asset (rehabilitation or replacement) is recorded and reported. In the private sector, a similar format reporting is required at the time of annual tax returns.

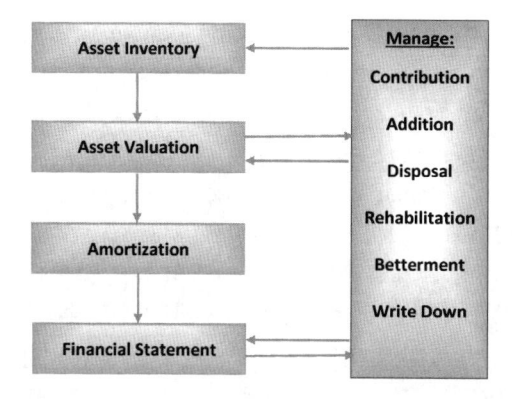

Fig. 10.5 Financial Reporting Process for Physical Assets

A good way to make the link between physical asset management and finance is to remember that deterioration decreases the condition of physical assets, and depreciation decreases their value. It is essential to balance deterioration with depreciation and condition with value.

Case Study 10.1

Gropalto Ltd has a total of $100M of physical assets, which are depreciated at 20% per year, straight line. The company is doing good business, utilizing its physical assets to full capacity and making annual profit of $10M every year. The return on investment (ROI) each year from those physical assets is shown in the Table 10.6 below:

	Year 1	Year 2	Year 3
Profit reported	$10M	$10M	$10M
Net book value of physical assets	$100M	$80M	$60M
ROI	10%	12.5%	16.67%

Table 10.6 — Physical Assets Return on Investment

The above shows very good ROI, which is increasing over the years. On paper, from a financial reporting perspective, it looks good. Managers may be reluctant to replace ageing physical assets with modern ones or to invest in additional physical assets, even though it could be beneficial in the long-term. From a financial strategy, this makes sense; even if we factored an increase in maintenance costs, the ROI would still be increasing. This scenario shows the gap that can exist in organizations when it comes to decision making to replace physical assets or invest in new physical assets.

We have assumed that the net book value of the physical assets is the sole component of investment in ROI calculation. In reality, the profit would be injected back into physical assets, both fixed and current.

From a physical asset management perspective, how will you deal with this situation? And what from a maintenance management perspective?

Going back once more to Fig. 4.8, we see that financial planning is involved in the planning and disposal phases of the physical asset lifecycle. On the planning side, we mentioned earlier that, if needs are not identified and defined in a timely manner, the right decisions cannot be made with regards to physical assets and hence will not be included in the capital plan. Condition assessment and needs assessment must be carried out on a regular basis (recommended every three years) to facilitate and support the decision-making process and to include those decisions in the capital plan.

On the disposal side, in many organizations nowadays, there is a huge disconnect between what is reported in the financial statement and what the organization currently has as inventory. This can be partly due to:
- Major gaps in the inventory updates
- Inability to keep track of physical asset changes within projects
- Non-existing or inconsistent physical asset valuation process
- Irregular reporting format to finance

The key questions for many organizations are the following. Who owns the physical asset (financial and technical) information? Who makes the link between the technical and financial information?

Currently, there is a huge gap between both since there is no ownership. The physical asset management function is the function that will bridge the gap to ensure physical asset inventory data and physical asset valuation numbers are consistent and are accurately recorded/reported.

CHAPTER 11 —
Physical Asset Management

*"The key is not to prioritize what's on your schedule,
but to schedule your priorities."*
~ Stephen Covey

Welcome back to the physical asset management function! After a long crusade exploring the different functional areas of organizations, and after having seen what they are all about, it is time now to look at the physical asset management functional area itself. We agree that there are G&O existing between functional areas in organizations, and we are confident that a holistic physical asset management approach can address those G&O for the benefit of organizations.

"What would a physical asset management function consist of?" you wonder. "Is it a profession, or just a collection of practices from different functions?"

These are interesting questions that I am sure have been in the back of many readers' and physical asset management practitioners' minds for a long time.

So far we have seen that physical asset management is a complex activity, to which there is no scientific approach. By this time you might well feel that we have not been able to pin down exactly what physical asset management is, but all is not lost. From Chapter 3 we saw that functional areas tend to regard business issues from their own perspective with focus on their ToF. We have also seen that each functional area plays a very distinct role in the lifecycle of physical assets with different and sometimes conflicting results. To be able to have a holistic approach to physical assets it is necessary to synthesize the functional areas. To complicate matters the fact that functional areas deal with complex issues mean very often they have to make critical decisions with respect to physical assets. It is never quite clear whether those decisions are satisfactory and to the benefit of the physical assets and the whole organization. Furthermore, there are so many internal and external factors that keep changing, with result that, maybe the optimum decision has not been made. This means that it is necessary to bring very high level of evaluation skills to analyze and choose among the different options. This brings us to two skills which are fundamental to a strategic approach to physical asset management.

Holistic physical asset management is a strategic endeavour in itself similar to the strategic thinking process as shown in Fig. 11.1 below.

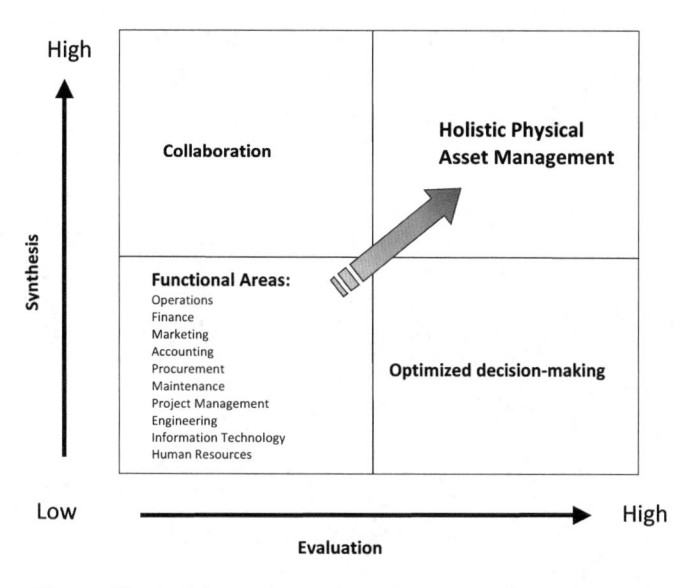

Fig 11.1 Physical Asset Management Strategic Holistic Approach

Functional areas are able to synthesize and evaluate in the bottom left corner within the context of their own functions. The problem is that a physical asset management problem cannot be resolved by the application of only one functional area in isolation. The physical asset management function's task is to bring these together through collaboration (moving up the vertical axis) and identify the best approach to tackle the particular issue. The physical asset management function then has to weigh up the pros and cons of potential alternatives and arrive at an optimum decision (moving along the horizontal axis). Thus the physical asset management function need to have a more holistic approach with a broader view than the individual functional area. The physical asset management function's task is not an easy one, especially to get to the top right hand box. The mandate of holistic physical asset management is to provide the structure within which all functional areas can be synthesized and all decisions are well evaluated.

Physical Asset Management Mandate

The mandate of the physical asset management function falls in three distinct areas within the physical asset lifecycle phases. Fig. 11.2 below depicts the typical bathtub curve of a physical asset with failure plotted against time. We can clearly see the infant mortality, the wear-out and the economic life phases of the physical asset. The three shaded areas are the mandate areas of the physical asset management function:
1. Planning
2. Decision-Making
3. Addressing G&O

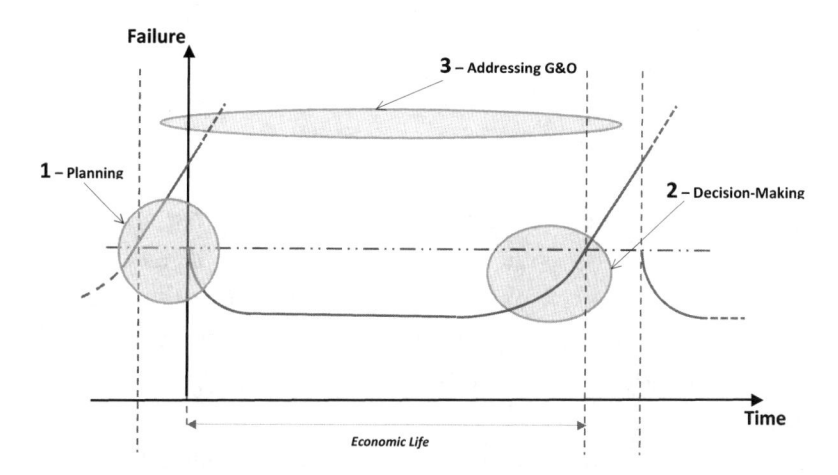

Fig. 11.2 Physical Asset Management Function Mandate

(1) PLANNING

The planning area is the period before the physical asset has even been put into operation. Notice how the shaded area "1-Planning" overlaps both the end and the beginning of the physical asset life. This planning area is similar to the asset planning phase, as described in Chapter 10 and consists of two key elements:
- Needs assessment
- Demand management

Needs assessment is a systematic process for determining and addressing needs, or "gaps" between current conditions and desired conditions or "wants." A needs assessment is a part of planning processes, often used for improvement in physical assets, processes, programs, organizations, or communities. It can be an effective tool to clarify problems and identify appropriate interventions or solutions. Particular methodologies that are found useful in needs assessment include economic appraisal, feasibility studies, value management, demand management, and risk management, among others. By clearly identifying the problem and analyzing the different options, finite resources can be directed towards developing and implementing a feasible and applicable solution to address the organization's needs and wants.

Demand management is about managing the demand side of the demand-versus-supply function. It is the practice of manipulating demand to ensure it meets the available supply. Demand management identifies and influences demand for assets. It utilizes techniques and non-asset options for delivering services by reducing the demand for new assets and hence saving money. Prior to injecting money in new capital physical assets, a needs analysis is carried out to assess the needs, the full lifecycle cost, and any other alternative solutions, to determine whether the demand can be altered. Demand management is an effective tool in influencing the need for physical assets and is an essential part of any effective physical asset management process.

The above two elements gives a good overall assessment of the situation, internal and external to the organization, to guide subsequent planning decisions and demand forecasts for future strategic planning. This may include installing a new equipment, expanding the facility, adding a new lane to the highway, building a new school in the area, etc.

(2) DECISION-MAKING

The decision-making area - shaded portion "2-Decision-Making" is located at the end of the physical asset economic life. It deals with the phase when the physical asset has been in operation for some time. A few key components are critical for developing a robust decision-making model for physical asset-intensive organizations:

- Condition assessment
- Service level assessment
- Risk assessment

Condition assessment is the process of performing an analysis of the condition of physical assets. Qualified professionals carry out a detailed walk-through inspection, which may be followed by further testing, when required. The outcome of a condition assessment is the collection of baseline data on the physical asset, including asset type, age, expected useful life, condition, estimated remaining life, valuation estimate, and replacement cost. Several condition rating systems can be used to assess the structural condition and functional adequacy of the physical assets, according to standard engineering practices.

Service level assessment is the process of evaluating the level of expectations that customers are looking for in certain goods or services. Service level is tied to technical level of service which defines the technical condition and performance of physical assets. Service levels, and to that effect, technical levels of service can be altered based on growth, new demands, changing needs, current condition of physical assets or even climate constraints. Understanding those changes will help make timely decisions on how to better provide the required levels of service.

Risk assessment is the process of assessing risks and utilizing the outcomes in decision-making. Often organisations will develop risk registers and develop processes to support the review and escalation of each potential risk..Organizations need to understand and agree on the risk tolerance level to assess the potential risks at the appropriate level of the organization. Proactively acting on the identified risks will help organizations prioritize and allocate the right resources to manage the consequences. ISO 31000, Risk Management Principles and Guidance is a good tool to apply good practices.

Knowing the condition of physical assets, the service levels required and the potential risks prevailing will provide good rationale and solid foundation for optimum decision making criteria. The reasons for making physical asset decisions could be due to operational requirements, financial implications, ageing assets, technological obsolescence, changes in regulations, or other business needs. For example if physical assets need replacement — do we go for like-for-like replacement or do we upgrade based on current and future needs. Is there any risk of obsolescence or potential of new upcoming technologies in near future? Techniques such as risk analysis, lifecycle costing, cost-benefit analysis, root-cause analysis, condition assessments, and many others are applied to combine technical, financial and non-financial data - all necessary to facilitate an optimized decision-making process. A significant amount of physical assets/operations data crunching and analysis for different alternatives takes place at this stage. Consultation with the different stakeholders is equally important so that informed decisions are made to meet the short-term and long-term organizational strategic objectives, while maintaining a good balance in service level, cost, and risk.

(3) ADDRESSING G&O

Addressing G&O is the third area of mandate of the physical asset management function - shaded area "3-Addressing G&O". It is the area extending over the whole lifecycle of the physical asset. We have seen, in the previous chapters that G&O exists between all the functional areas involved in one way or another with physical assets. This mandate of the physical asset management function is mainly concerned with building synergies and tight working relationships among the different functional areas to *make the holistic approach a reality*. Harmonized interactions need to happen consistently to proactively assess physical asset performance level, service level, risk, and cost, and to ensure that the right informed decisions are made at the right times. In industry terms, this means the physical asset management function is tasked to bridge gaps between functional areas so that the organization reaps maximum value from its physical assets.

For the physical asset management function to successfully fulfill its mandates in terms of proper planning, timely and optimized decision-making, while applying a holistic approach, it needs to work in harmony with each functional area.

Let us take a look now at how that harmony can be realized.

Harmonization with Project Management

One of the key elements for the success of the physical asset management in an organization is to address the G&O with the project management function. This is the start of everything and, if we get it right at this stage, things can be much better over the entire lifecycle of the physical asset. The working relationship is of utmost importance to ensure that the right physical assets are designed and delivered to meet identified needs in a timely manner. From the moment physical asset management hands over a business case or physical asset need to the project management function, to the execution and handing over of the project to operations and maintenance, many things can happen. First of all, there should be a smooth transition of the business case or physical asset need to the project management function.

- How and in what shape and form does the transition happen?
- Is it a business appraisal, a pre-feasibility study, or a needs-assessment report?
- What other information or data is required—budget estimates, scope of work, timelines?
- What are the risks and constraints from an organizational, program level and physical asset perspective?

Once handing over is completed, the project team needs to translate all this into a detailed design and executable package, while ensuring the required scope, timeline, and budget are kept and met. During the whole process, both functional areas need to work together to make sure configuration management, value engineering, system engineering, reliability, and maintainability principles are included.

"What if anything changes during project execution?" you question.

Well, obviously, as mentioned earlier, many things can happen during a project lifecycle. To make sure nothing is left out or things do not fall behind, all changes must be reviewed and signed off. A process with check points needs to be in place to validate all changes that may have an impact on the physical asset function, performance standard, RAM, lifecycle cost, and other specific requirements.

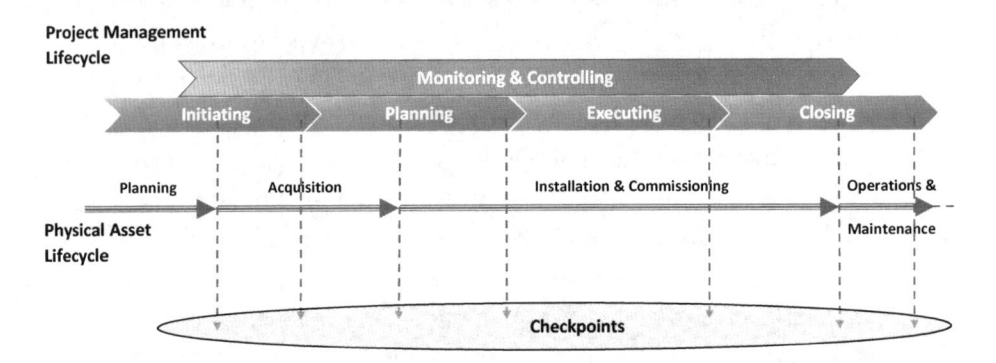

Fig. 11.3 Check point process between physical asset management and project management functions

Fig. 11.3 above shows this process. It is important that the project management and physical asset management functions, or whoever executes capital projects in the organization for that matter, adhere to this process.

Harmonization with Maintenance Management & Reliability Engineering

Other functions of utmost importance to the physical asset management function are the maintenance management and reliability engineering functions. These functions must interact closely with each other to address all G&O and to ensure that, at the design phase, they are in compliance with reliability and maintainability principles. Also crucial are the commissioning and handing over phases. The physical asset management function has to ensure that physical asset knowledge and data are transferred to the organization, and to the other functional areas. The physical asset register must be updated in the asset management system, as well as in the CMMS, with all related current physical asset information and documentation. The physical asset management and maintenance & reliability functions must ensure that maintenance and reliability practices are developed and loaded in the work management system. This is very important to ensure physical asset performance standards are maintained during operation, risks are minimized, costs are contained, and, of course, the original expected physical asset life is preserved.

Over their entire lifecycle, physical assets can fail in many different ways and for many reasons. Some of the main reasons are:
- Normal wear and tear
- Age
- Insufficient or lack of proactive maintenance
- Poor operations
- Random failures
- Unforeseen accidents

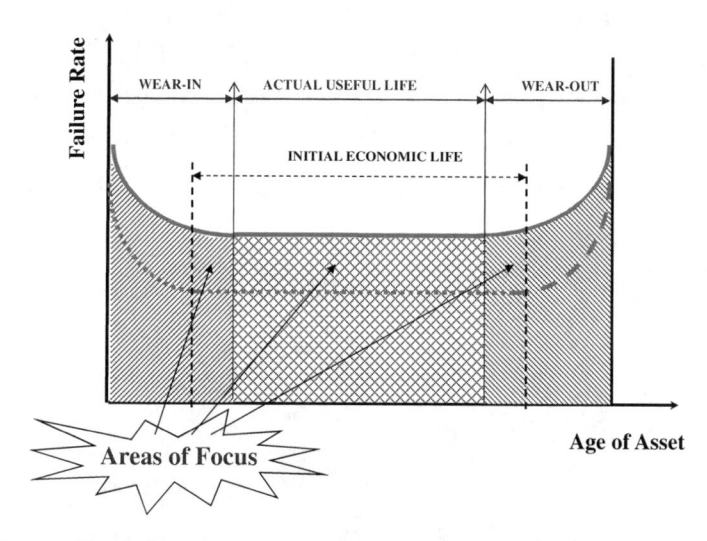

Fig. 11.4 Physical asset management, maintenance management & reliability engineering focus areas

Fig. 11.4 above shows the areas of focus of the physical asset management, maintenance management, and reliability engineering functions to ensure the failure rate curve remains as low as possible to maximize economic/useful life. They must work together to ensure the wear-in, random and age-related failures are minimized as much as possible, so that the initial economic lives of physical assets are preserved all the way. This will help physical asset management do better planning, be more proactive, and make the right decisions at the right times.

"From what I am seeing, it looks like the project management and the maintenance & reliability functions are critical for the success of physical asset management function," you say.

In a sense, yes. These two functions, as you now know, work very closely with physical assets and they do so at very critical phases of the lifecycles. So it is obvious that project management and maintenance & reliability will be the main allies of the physical asset management function. There has to be constant communication, back and forth interaction and exchange of information among the three functions at all times. That synergy, as shown in Fig. 11.5, is necessary to ensure the original decisions made about physical assets are still aligned with the strategic objectives of the organization. It is a top-down, bottom-up approach, with forward-looking plans, based on current conditions.

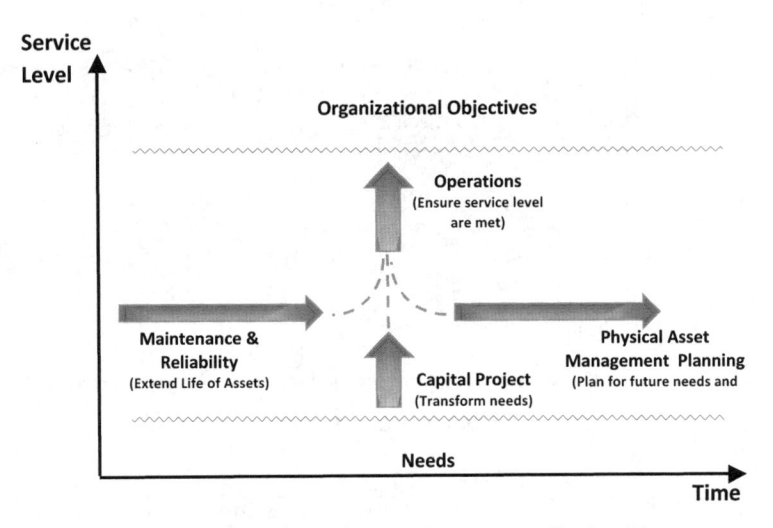

Fig. 11.5 Synergy between Physical Asset Management, Project Management, and Maintenance & Reliability

Harmonization with financial planning

Finance is also very critical to the success of the physical asset management function. No matter how well the physical asset management function lines up its asset plan and its capital plan, if finance is not on board, no plan can be materialized. Physical asset management should work closely and interact with the finance planning function to address all G&O, not only during the capital budgeting period, but throughout the whole year, on a consistent basis. The main goal is to record and to relay important information and decisions related to physical asset capital investment that may have impact on the finance plan and cash flow, such as:
- The short-term and long-term needs of operations
- Physical asset related issues and feedback from maintenance and reliability
- Any changes encountered by project management in their projects with respect to scope, time, and cost
- The future needs of existing physical assets, based on their current condition
- Any other physical-asset-related risks to the organization that need action

The documentation and investigation of physical assets and operational needs are critical for asset and capital planning. Without proper asset and capital planning, sound financial planning is impossible.

Like the saying goes, "An asset/capital plan without a financial plan is a wish list, and a financial plan without an asset/capital plan is a waste list!"

And of course in all that, time is of the essence.

We all know that, with time, the condition of physical assets deteriorates and, if the rate of deterioration is not controlled and managed, it can be challenging to make the right decisions and come up with effective short-term and long-term asset and capital plans.

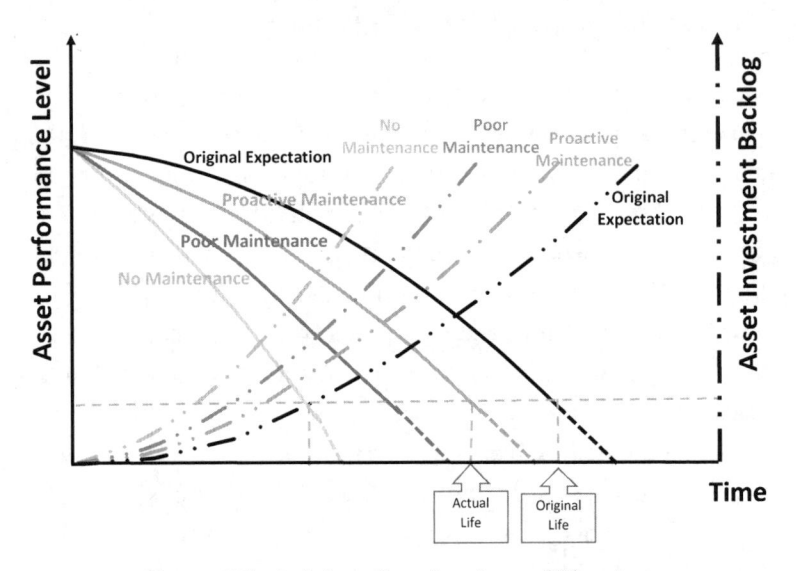

Fig. 11.6 Physical Asset Deterioration and Planning

Fig. 11.6 (adapted from *Danger Ahead: The coming collapse of Canada's Municipal Infrastructure, Prof. Saeed Mirza, November 2007*) shows the effect of maintenance on the life expectancy of physical assets and the impact on investment backlog. All physical assets have a certain expected service life. This service life will fluctuate based on the level of maintenance performed on them. With no maintenance at all, the actual service life will be considerably short, while with proactive maintenance actual service life can be very close to the original service life. No or poor maintenance will accelerate the rehabilitation/replacement frequencies of physical assets, which will add up faster to the investment backlog. So it is important for everyone in the organization to do their part so we know where we stand and can plan accordingly. Poor planning leads to reactive situations, resulting in huge outlays of funds and occasionally the bypassing of policies/regulations to respond to emergency situations. Reactive situations can have impact on financing strategies and future planning. That is why the success of a physical asset management function, in many organizations, is very much dependent on the effectiveness and accuracy of financial planning and vice-versa.

Case Study 11.1

Operations at Sirrokane Ltd. requested that the main mixer tank be relined before their peak production period. The need for the refurbishment task was approved by physical asset management and planned for the next year, with capital funds set aside. The scope of work was outlined in a need assessment with a budget estimate, and timeline for execution. The project was assigned to the project management group. The project management group worked on the project for some months with a consultant and came back with a project plan. As per the project plan, the cost was almost twice the initial estimate, with a significantly different scope of work and the project was forecasted to take way more time than the requested timeline.

Operations was not too impressed because, on one side, they could not shut production down beyond the originally requested timeline and, on the other hand, they were worried the mixer tank will not be able to last through the peak production period.

Maintenance also was not pleased because the tank was leaking in a few places and they had to go in the mixer tank an average of three times every week to do some after-hours, temporary repairs. This was taking a toll on maintenance's workload and budget. Financial planning on the other hand was complaining that they do not have sufficient funds and they asked that other project(s) be postponed to free up funds.

The VP of Operations called a meeting with all five functional areas—Operations, Finance, Physical Asset Management, Project Management, and Maintenance—to discuss the situation, find out what the issues are and to come up with a solution.

The case study clearly demonstrates the need for functional areas to keep an open line of communication at all times, to understand

each other's needs/priorities, and to take proactive decisions. A few key points to highlight:
- Were the initial scope of work, budget and timeline accurate?
- Did they meet operational needs? Short-term and long-term?
- Did we clearly delineate needs from wants when the consultant worked on the design?
- When cost changes, it may impact the funding strategy and, potentially, the business case. We may want to revisit the original decision.
- When the timeline is not met, it can impact operations and increase the risk to the business.
- When the timeline is not met, it can weigh heavily on the maintenance team, with more reactive work.
- At the end of the day, physical assets continue to deteriorate, and any remedial action must be undertaken in a timely manner.
- The more we wait, the more the scope of work may change due to changes in physical asset condition.

The above illustrates some of the reality existing in many organizations, where there are huge gaps in communication and decision making, especially when many functional areas are involved working in silos. Physical asset management function must provide that wider approach and bring everyone & everything together.

"What about the other functional areas?" you ask. "Aren't they important?"

You mean, besides project management, finance and maintenance & reliability? Sure, they are all equally important! For example if we throw in Procurement we will uncover other challenges. Each functional area is as important as the other to create the holistic physical asset management approach. Try to picture the gear train in a watch: the movement of the wheel train is all in sync. Similarly, in an organization, each functional area involved with physical assets has to do its part to make the holistic physical asset management clock tick.

Think of the situation this way. Let us assume the organization is the ring gear of a planetary gear system for physical assets, with operations, project management, and maintenance as the planet gears, as shown in Fig. 11.7.

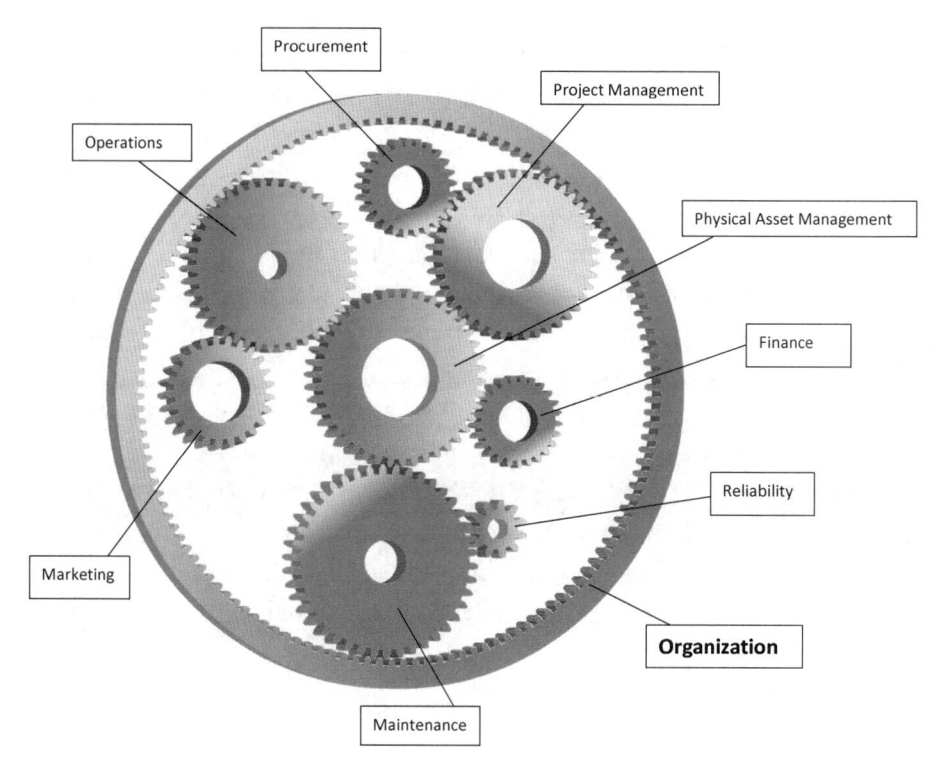

Fig. 11.7 Physical Asset Management Planetary Gear System

All the functional areas work hard to rotate clockwise and make the organization rotate in the same direction. We know also that operations works closely with marketing, project management with procurement, and maintenance with reliability. These functional areas are the satellites of their respective planets. Notice how the planet gears are all a distance away from each other, and that bringing the three planet gears together will not work. (*Imagine for a moment the three gears Operations, Project Management and Maintenance, touching each other — for sure there will be a direction of rotation problem*).

Now if the speeds of the three planets are not synchronized, it may affect the speed of the planets overall (one will slow down and the other will speed ups depending which one is the driver!). Ultimately it will affect the speed of the ring gear — the organization.

Consider the situation with a sun gear in the middle, bringing every planet and the ring gear into synchronization. The sun gear in the physical asset planetary gear system represents the physical asset management function, which harmonizes the speed of the planets so the organization can function well.

"But what about the gear next to the sun?" you ask.

Well seen. The gear or satellite of the sun is the finance function. It is the good friend of physical asset management, providing the required funds and cash flow based on the agreed upon physical asset and capital plans.

Using the above analogy, we all recognize the need for better interaction, collaboration, and synchronization between all functional areas. The physical asset management function plays a central and pivotal role in physical asset management for an organization to bring harmony in that interaction and to help all functional areas work out the best solution to achieve organizational objectives.

CHAPTER 12 —
Implementing Physical Asset Management

"There is nothing more difficult to plan, more doubtful of success, nor more dangerous to manage than the creation of a new system. For the initiator has the enmity of all who would profit by preservation of the old system and merely lukewarm defenders in those who would gain by the new one."
~Niccolo Machiavelli

We have finished reviewing the pivotal role that a physical asset management function plays in an organization. There is no doubt that physical asset management is an emerging discipline, prompted by global trends and future uncertainties, such as ageing infrastructure, economic recession, population growth, environmental challenges, energy crisis, demographic changes, and climate change. Organizations, whether in the public or private sector, need to be aware of the situation and be ready for a more proactive, integrated, optimized, and risk-based holistic physical asset management approach. If you look at the public sector, alone, there is a huge physical asset (infrastructure) gap at all levels of governments.

In Canada, for example, spending in infrastructure has fallen behind by $123 billion since the 1950s (*2015 Canadian*

Infrastructure Report Card), while in the United States an estimated $3.6 trillion investment will be needed by 2020 (*2013 Report Card for America's Infrastructure*). If nothing is done, not only will the deficit grow bigger, but productivity at the country level will be adversely impacted. World-class physical asset is the backbone of any country's economic productivity. Better roads, transportation, communication infrastructure, utilities, and water systems result in lower business costs and increased profits. So, on one side, organizations are ready to take on more debt to provide the funds required to invest in physical assets and, on the other side, organizations are facing the dilemma of where to invest their money. Should it be on projects that will achieve long-term productivity gains, such as better public transit, or should it be on the so-called "shovel-ready" projects, such as upgrading the local community center, or should it be on long-term benefits projects, such as sewer upgrades? How will organizations approach this situation and make sure funds are spent wisely, on the right physical assets, in a timely manner?

Faced with all these tough questions, organizations are coming up with innovative approaches in the form of physical asset management frameworks to bring all the pieces of the puzzle together to achieve excellence.

The physical asset management function, together with other functional areas, fulfills the mandate of balancing cost, service level, and risk via physical asset lifecycle delivery strategies and optimized decision-making processes. In the previous chapters, we have seen that many organizations are struggling to make that happen due to several fundamental challenges, as summarized below:

- Fitting the physical asset management philosophy in the organizational structure
- Coping with the intricacies of interaction between different functional areas
- Addressing the existing gaps and overlaps over the whole physical asset lifecycle
- Sustaining the paradigm shift

"These look quite important for the success of physical asset management," you add. "How should organizations cope with them?"

Well, organizations must definitely overcome these challenges. If you look at Fig. 12.1, these challenges represent the foundation and building blocks of the holistic physical asset management house of excellence required to support and achieve the service level, risk and cost equilibrium.

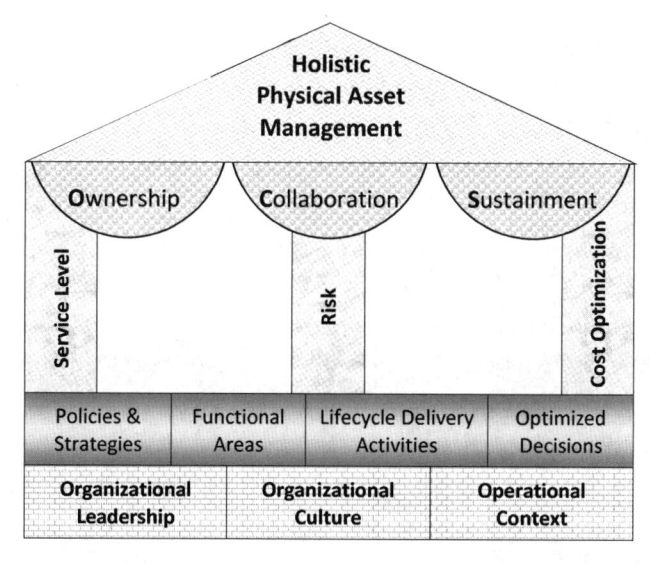

Fig. 12.1 The Holistic Physical Asset Management House of Excellence

Implementing a successful holistic physical asset management philosophy in an organization requires that these foundational challenges are addressed. To guide organizations in their implementation journey, we propose a three-step approach:
- Step 1 — Ownership
- Step 2 — Collaboration
- Step 3 — Sustainment

The three-step approach will be thereafter called the OCS Approach. The OCS Approach will help organizations transition from their current state vis-à-vis physical asset management to a holistic physical asset management state, while keeping the balance of service level, cost, and risk. The transformation of the business will take place via physical asset management policies & strategies across all functional areas, via lifecycle delivery activities, and via optimum decision-making processes. So let us take a look at the three steps of the OCS Approach.

Ownership

In previous chapters, we have seen that there are different functional areas dealing with physical assets at different points in time over the entire physical asset lifecycle (remember the "hot potato" in Chapter 4). The first big question that needs to be answered in the OCS Approach is:
"Who owns the physical assets in an organization?"
I bet, if you ask this question around in your organization, you will get different answers. Some may say it is operations. Others will say, "No, it's maintenance. Yet others may say, "No, it's actually finance."

They are all correct in some ways because there is, in fact, no real ownership of physical assets in organizations—it is a shared **ownership**. Maintenance will own the physical assets when they are doing maintenance or repairs on them; operations will own the physical assets when they are producing some kind of output; and, project management will own the physical assets when they are upgrading them.

Physical asset management is the function that would provide the real ownership of physical assets, throughout all the lifecycle phases, for the benefit of the organization. This does not mean that physical asset management will do or be responsible for all the work that needs to be done on the physical asset, but rather it will be involved in decision making, funding, and timing to support the other functions. Physical asset management's main purpose is to own the physical assets on behalf of the organization and make sure all decisions made are in line with organizational goals and objectives to maximize value.

"How does an organization bestow that ownership to the physical asset management function?" you ask right away.

I knew this question was coming because it is a logical one to ask after what we have seen above. Well, the question to the question of assigning ownership is: "Do we need a major reorganization to create a separate physical asset management department or do we leverage the existing functional departments and cross-functional processes?"

My answer to this is: Go with whatever works and makes sense to the organization. At the end of the day, what matters is that ownership of the physical assets is clarified and agreed upon. Of course, the ideal situation is to create a physical asset management function wherever possible, as many organizations are already doing, with dedicated staff attached to it. Maybe the physical asset management function will be required at first to set up the framework and then eventually phase out. No matter what, the space for physical asset management and, to that effect, for physical asset ownership, has to be created, be it in an existing functional area or as a separate dedicated function.

"What is the ideal organizational structure then?" you ask.

There is no ideal model or structure, but there are certain key factors that need to be considered when you try to fit a physical asset management function into an organization:
- The industry and applicable regulations
- Prevailing organizational culture
- Current business priorities
- Physical asset ownership and decision-making process
- Single-point accountability
- Number of physical asset types and challenges

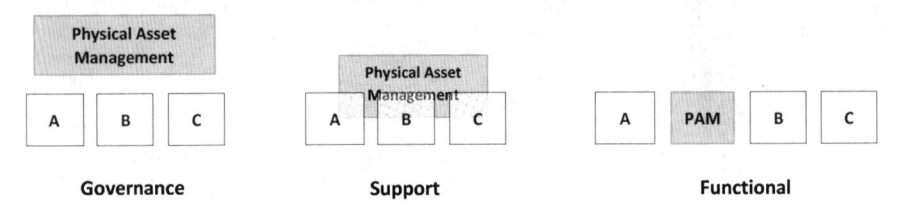

Fig. 12.2 Physical Asset Management Fit into Organizational Structure

Fig. 12.2 above depicts three potential organizational structures to fit the physical asset management function:

1. **Governance**: In this organizational structure, the physical asset management function is placed at the corporate level. Its main mandate is to drive physical asset management from the top, and it acts mainly as a facilitator to the various business programs and other functional areas. The governance model develops policies and strategies to standardize practices across the whole organization, while implementation is left to the business programs.
2. **Support**: Here the physical asset management function is embedded in the organization and provides support to all business programs and functional areas. The physical asset management function reports to the higher hierarchy via another reporting level. Its mandate is to develop and implement policies and strategies, while facilitating tactical implementations with the business programs.
3. **Functional**: In this structure, physical asset management is at the same level as the other functional areas. It forms part of the operational group and has its own mandate to deliver. It is responsible for developing and implementing all strategic and tactical projects for the physical asset management function.

Ideally the physical asset management function should operate mainly as a decision-maker to support the other functional areas. It is not desirable to see the physical asset management function leaning towards a strictly policing role or to a doer role. It has to have the right balance in the organizational structure for functional areas to reach out to them.

One key point to note here is to whom the physical asset management function will report to in the organizational hierarchy. Issues with regards to conflict interest can arise when it comes to making critical decisions. It is ideal for the physical asset management function to be unbiased when making those decisions, which should be based solely on risk assessment, level of service, and cost optimization, to the benefit of the organization. The decisions should not be based on the wishes of project management, or maintenance, or even be influenced by operations. The physical asset management function is better placed to have a holistic view of the situation and to be the watchdog to separate needs and wants. The crucial idea here is to preserve the neutrality of the physical asset management function and to work closely in partnership with the other existing functional areas.

Collaboration

Now let us move on to the second step of the OCS Approach, which is **collaboration**. It is a known fact that no organization can function without collaboration. There is a management revolution taking place in the business world, with a new kind of organization emerging that is capable of achieving both continuous innovation and transformation, along with disciplined execution. To implement a whole-lifecycle physical asset management philosophy, such organizational transformation is required to break free from the classic command-and-control leadership style and organizational culture, to move toward a collaborative, connected, and participative model. Collaboration is, indeed, a top priority for many organizations and their functional areas. After all, no two functional areas are alike, and their strategies and expertise can be quite different.

166 Physical Asset Management: An Organizational Challenge

Collaborative initiatives come from different functional areas, with different ToF, different approaches, goals, and measures of success.

"So, with so much variety, how do we get them to collaborate to achieve the common goals and objectives of an organization?" you question.

The answer can be quite complex and relies on the organization's leadership to engage their personnel via connected leadership, pervasive learning, and collaborative practices.

The approach by organizations to adopt collaborative practices is represented by the twelve habits of highly collaborative organizations, as depicted in Fig. 12.3.

12 PRINCIPLES OF COLLABORATION

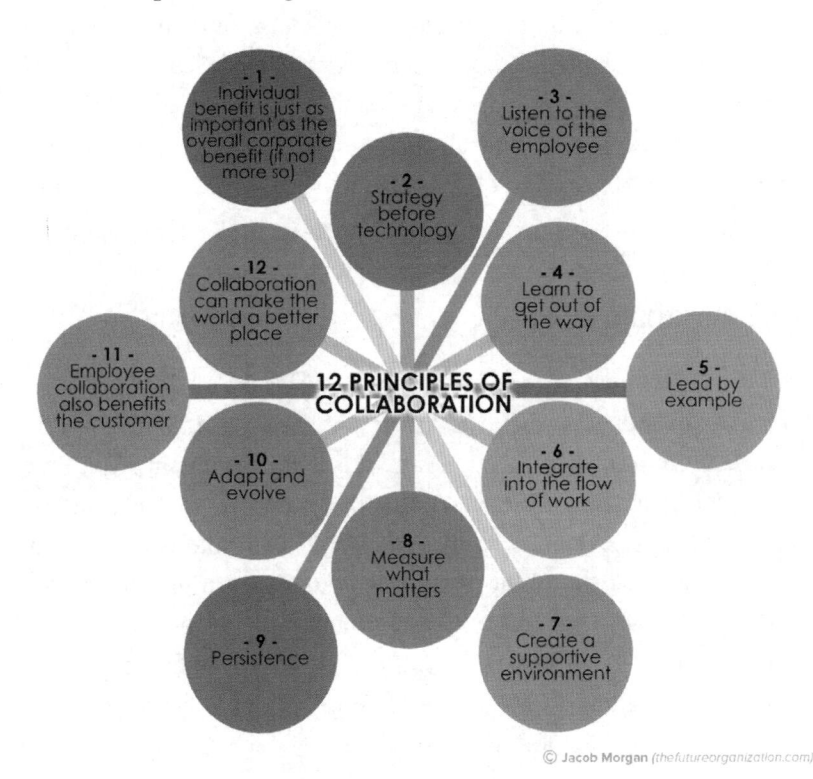

- 1 - Individual benefit is just as important as the overall corporate benefit (if not more so)
- 2 - Strategy before technology
- 3 - Listen to the voice of the employee
- 4 - Learn to get out of the way
- 5 - Lead by example
- 6 - Integrate into the flow of work
- 7 - Create a supportive environment
- 8 - Measure what matters
- 9 - Persistence
- 10 - Adapt and evolve
- 11 - Employee collaboration also benefits the customer
- 12 - Collaboration can make the world a better place

© Jacob Morgan (thefutureorganization.com) & the FOW

Fig. 12.3 Twelve Habits of Highly Collaborative Organizations
-Jacob Morgan Chess Media Group

To implement a holistic physical asset management, we need the full collaboration of many functional areas of the organization. It is obvious that a physical asset management function cannot exist and operate on its own. On the contrary, one of the main mandates of a physical asset management function is to break existing silos and foster collaboration across different functional areas. In short collaboration defeats functional silos.

For example, we have seen that maintenance and reliability, project management, finance, and physical asset management need to work together, over time, to transform needs into organizational objectives. Physical asset management will identify and approve the needs. Finance will put in place funding strategies so that long-term needs, identified by the physical asset management function, are achievable. The project management function will ensure that the transformation of needs into service level happens in a timely manner. Maintenance and reliability will take care of the short-term physical asset lifecycle delivery activities by doing routine maintenance and minor repairs. All these actions must be coordinated accordingly over time, and require intensive collaborative efforts and disciplined execution from each functional area so that cost is optimized, risk is minimized, and required service level is maintained. In effect, many of the twelve principles of collaboration (if not all) can be applied here. The collaboration efforts must happen right at the beginning so that we don't hear the common "I was not involved" or "they did not check with us before going ahead" that we experience in many situations.

"So, where do we start the collaboration effort?" you ask.

Everyone in the organization must put their heads together to come up with a solution to start collaborating, as shown in Fig. 12.4 below. The "face" epitomizes the collaborative efforts and how functional areas can work together to achieve holistic physical asset management.

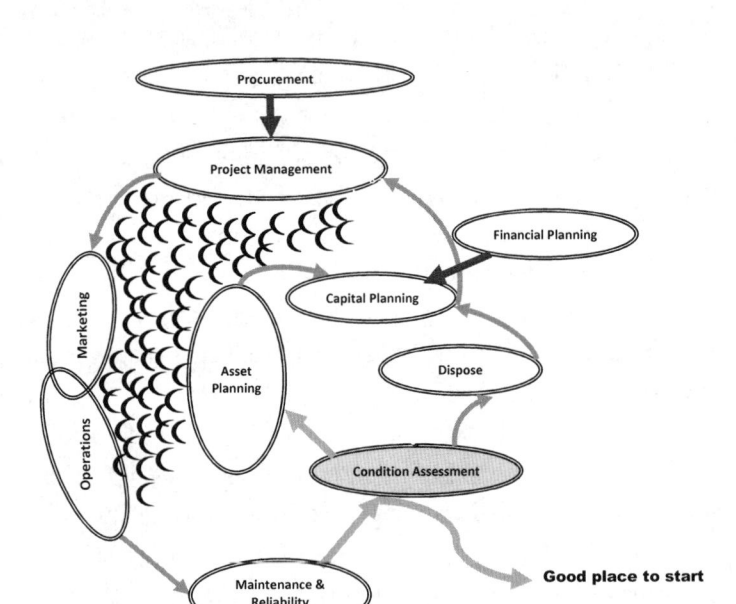

Fig. 12.4 Holistic Physical Asset Management Collaborative Efforts

One good place to start the physical asset management collaborative efforts with other functional areas is to carry out a condition assessment to have a 'taste' on what's coming up. The physical asset management function can initiate this effort, link up with all the other functional areas, to come up with a plan and act on the recommendations. There can be a 'feeling' to dispose of the physical asset or maybe it will 'sound' better to plan for future needs and growth. Capital planning will take a closer 'look' at what is needed in the short-term and plan execution. Financial planning will continue to 'look further' in the future for multi-year budget forecast. Project management will 'think' the ideas through, put all together and make it happen with the 'blessing and support' from procurement. All this is to provide operations and marketing with a strong 'back bone' to carry out their business while maintenance & reliability will ensure a solid and stable support.

Sustainment

The third and last step of the OCS Approach is *sustainment*. No matter how your organization decides to move forward in creating a physical asset management function, it needs to do so with the end state in mind for long-term sustainability. Thus, organizations should not jump on the physical asset management bandwagon and spend hundreds thousands of dollars without having a road map in place.

What could hinder the collaborative effort between functional areas during physical asset management implementation is the existing G&O. During our analysis of the main functions of an organization, we have seen that there are a lot of G&O when dealing with physical assets throughout their lifecycle.

To make sure the collaborative effort undertaken by functional areas does not go to waste, it is imperative that organizations address the G&O and streamline the business processes. To tackle the G&O issues, each functional area must be involved, because it may require some significant business process reengineering. Functional areas must be ready for the change and must manage the change in the best possible way. At the same time, they need to ensure that their own business needs and ToF are preserved and taken care of at all times. Support from senior management and inspiring leadership at all levels is important to influence, drive, and manage the change. New business processes will be developed and initiated; and old business processes will be revised, or even scrapped and replaced by new ones. All this can be overwhelming to some personnel and would need close attention from the different managers of the functional areas.

Business process reengineering with unique goal to address the G&O may bring with it some redistribution of roles and responsibilities. Some new and revised business processes will be allocated to the new physical asset management function, while other functional areas may find themselves with some business processes taken away from them, and assigned with other new or revised business processes. For long-lasting collaboration, all functional areas must buy in to the new way of conducting business and must be comfortable with the change to align with the new direction the organization is taking.

"This sounds like a major change for an organization and, more specifically, for the functional areas. Don't you think?" you ask.

Not really. It will depend on the organization's size, structure, complexity, and physical asset-intensiveness. It is very important for organizations to be ready before they embark on the journey of physical asset management.

"What do you mean by ready?" you ask.

Ready means the level of maturity of the process and culture in place in the organization. Is the foundation in place? Are people ready to change their old habits? Are they receptive to new ideas, new ways of doing things? All this determines the level of maturity of the organization and gives an indication of its readiness to make the paradigm shift.

However, I agree that, in some cases, it might bring major upheaval depending on the maturity level of physical asset management prevailing in the organization at that time.

For an effective physical asset management system, organizations need to have a clear road map for how they want to move forward within their own operational context, based on the organization's level of maturity.

"What if senior leadership and top management do not buy in the concept and support?" you ask.

To do things right, and in a sustainable manner, first and foremost you have to have the buy-in, support, and leadership of top management. Only then you can proceed with putting together a strategic plan for physical asset management which aligns and links with the organizational strategic plan. A sustainable physical asset management implementation must comprise of the following key elements:

1. Physical asset management system

2. Physical asset management framework
3. Functional roles & responsibilities
4. Physical asset management structure
5. Physical asset knowledge base
6. Physical asset management information system

Physical Asset Management System

There are several leading physical asset management standards that are used to guide the principles and practices:
- *InfraGuide* (National Guide to Sustainable Municipal Infrastructure, 2004)
- *PAS 55* (British Standards Institution, 2008)
- *ISO 55000* (International Standards Organization, 2014)
- *International Infrastructure Management Manual* (National Asset Management Steering Group Ltd., 2011)
- *Ministry of Infrastructure Guide for Municipal Asset Management Plans* (2012 Canada)

To proceed with implementing whole-lifecycle physical asset management in your organization, a good start is to review the ISO 55001 requirements and follow the ISO 55002 guidelines to build a sustainable physical asset management system. As per ISO 55000, an **asset management system** is:

> A set of interrelated or interacting elements of an organization, that establish asset management policies and objectives, and the processes needed to achieve those objectives.

A physical asset management system in an organization will take care of all the challenges that we have discussed so far, will cut across the functional areas of an organization, and will foster harmonized interactions.

The physical asset management system requirements in ISO 55001 have been organized into seven specific elements:

1. Organizational context
2. Leadership
3. Planning
4. Support
5. Operation
6. Performance evaluation
7. Improvement

It is strongly recommended that organizations interested to implement a physical asset management system to review and follow the ISO 55000 suite of standards for guidance on the application of the above seven elements.

Physical Asset Management Framework

A key challenge for organizations starting to implement a holistic physical asset management philosophy is establishing a linkage between the outcomes of the physical asset management function and the organizational strategic objectives. A *physical asset management framework* outlines the physical asset management strategies, policies, standard processes, and performance measures that are aligned with the strategic objectives.

It includes a statement of the overall purpose of physical asset management as a new approach for managing physical assets, the mandates and basic types of activities involved in physical asset management and identification of the different stakeholders associated with the discipline in the current operational context. The purpose of the framework is to establish a foundation on which any activity associated with physical asset management across the organization can relate to a common set of operating principles to achieve an overall common organizational objective.

In reviewing and adopting the framework, it is important that stakeholders are able to agree with the statements that are made and to recognize the relationship their program or functional areas has within the overall physical asset management principles. The acknowledgement and endorsement of the physical asset management framework is an important step in moving towards integrating the collective efforts of the whole organization to address common strategic objectives

The physical asset management framework is intended to demonstrate how the individual elements fit together and how to formulate the decision-making processes to form an integrated management strategy.

The key elements of a physical asset management framework are:
- Organizational strategic goals and objectives
- Physical asset management policies that include level of service, performance measures, and risk management
- Physical asset management strategies that include asset level of service and performance measures, asset lifecycle strategies, maintenance plan, and capital plan
- Optimized physical asset- decision making practices, which include lifecycle costing and demand management
- A financial plan

Different organizations will have different models of physical asset management framework and may consist of more elements, but the main aim is to link the organizational, physical asset management, and operational objectives. In other words, the physical asset management framework consists of outlining and linking the strategic, tactical, and operational plans via physical asset management and financial plans.

Implementing a physical asset management framework is a challenge for many organizations. A successful implementation should follow a Plan-Do-Check-Act cycle (PDCA cycle) and would consist of the following:
- Carry out gap analysis to identify areas for improvement
- Prioritize recommendations and tasks
- Develop an implementation plan
- Track progress of implementation
- Assess physical asset management maturity level

- Repeat process for continuous improvement

The whole endeavour must have full backing from top management and clear support from the other functional areas of the organization. The physical asset management function, on the other hand, needs to be fully functional with clear direction and mandate.

Functional Roles & Responsibilities

To get a better idea of how much business process reengineering may be required when implementing a physical asset management function, let us take a closer look at the roles and responsibilities of functional areas in relation to physical asset management. For the purpose of this exercise, we will use the thirty-nine landscape subject groups of the GFMAM Landscape subjects. (The Asset Management Landscape, Second Edition, March 2014; Asset Management–An Anatomy, Version 3, December 2015).

These thirty-nine subjects describe the complete scope of physical asset management. Therefore, any organization planning to embark on the journey of whole lifecycle physical asset management must ensure that it has the necessary competencies, knowledge, and expertise in-house to focus on those subject areas. In Table 12.5 below we have attempted to assign each physical asset management landscape subject to a particular functional area, based on its best fit and area of expertise. The functional area will be the lead and will be responsible for the assigned subject. Of course, there will be other functional areas involved in each subject area to collaborate, but the roles and responsibilities will be clear as to who owns and who should deliver on each landscape subject. As you will notice in Table 12.5, it was not an easy task to assign ownership to different functional areas due to the existing G&O in many organizations. In fact the challenge starts and resides here — clarifying roles and responsibilities, ownership and accountabilities! You will also notice I have assigned "ALL" for some subjects. This means leadership and ownership should come from senior management. It goes without saying that the table will vary from organization to organization, and from industry to industry. Table 12.5 may not represent what you currently have in your organization, but I recommend you use it to map subject areas within your own organization, and functional areas based on your own operational context.

Subject Group	Asset Management Subject	Functional Lead
Asset Management Strategy and Planning	Asset Management Policy	AM
	Asset Management Strategy	AM
	Demand Analysis	AM
	Strategic Planning	ALL
	Asset Management Planning	AM
Asset Management Decision-Making	Capital Investment Decision-Making	ALL
	Operations & Maintenance Decision-Making	Maintenance
	Whole-life Cost & Value Optimisation	AM
	Resourcing Strategy & Optimisation	ALL
	Shutdowns & Outage Strategy & Optimisation	Maintenance
Lifecycle Delivery Activities	Technical Standards & Legislation	Reliability
	Asset Creation & Commissioning	Project
	Systems Engineering	Reliability
	Configuration Management	AM
	Maintenance Delivery	Maintenance
	Reliability Engineering	Reliability
	Asset Operations	Operations
	Resource Management	ALL
	Shutdown & Outage Management	Maintenance
	Fault & Incident Response	Reliability
	Asset Decommissioning and Disposal	AM
Asset Information	Asset Information Strategy	AM
	Asset Knowledge Standards	AM
	Asset Information Systems	AM
	Data & Information Management	AM
Organization & People	Procurement & Supply Chain Management	Procurement
	Asset Management Leadership	AM
	Organizational Structure	ALL
	Organizational Culture	ALL
	Competence Management	HR
Risk & Review	Criticality, Risk Assessment and Management	Reliability
	Contingency Planning & Resiliency Analysis	AM
	Sustainable Development	AM
	Management of Change	ALL
	Asset Performance & Health Monitoring	Reliability
	Asset Management System Monitoring	AM
	Management Review, Audit & Assurance	AM
	Asset Costing & Valuation	Finance
	Stakeholder Engagement	AM

Table 12.5 — Physical Asset Management Landscape Subject and Functional Areas

Physical Asset Management Structure

Earlier we saw that organizations can structure themselves differently to fit the physical asset management function. Now the question is: what would be the best physical asset management function structure and who should be on the team in terms of knowledge and competencies? Let us start with the physical asset manager. Do we need one or not? This will depend on the operational context, size, complexity, and many other factors. What is important is to have the correct, unbiased leadership in place to make the right decisions for the good of the organization.

"What would be the role of an Asset Manager?" you ask.

The *asset manager* will have to be involved in the development of the physical asset management framework, which will include the physical asset management policy and strategy. He or she will work with senior management and the different program areas to ensure the link with the organizational strategy. The asset manager will also be required to work closely with the finance department and be involved in the capital planning process. He or she will be looking at the different options for activities and investments going forward, to come up with a plan fully aligned with the strategy. The asset manager will be working with the maintenance and project management functions to ensure all physical asset lifecycle delivery activities are completed effectively and efficiently. Another key role for the asset manager is to assess and manage risk of action or inaction on the performance of physical assets, in the context of the organization's strategic objectives. Lastly, and most importantly, the asset manager is required to collect and collate all the pertinent data and information with regards to physical assets so as to facilitate and support all decisions.

"That's quite a variety of skill sets for one individual, isn't it?"

Sure it is. And I must admit that it has to be, so as to be able to effectively lead the physical asset management team as well as support the organization. An asset manager should not only have sound technical knowledge with years of working experience with physical assets, but must also have good business administration knowledge with a strong financial background. The balance is very important for this key role to better understand the implications of physical asset decision-making, backed with financial and organizational impact.

Going back to the physical asset management team, we know that it has to work closely and collaborate with the different functional areas. The team needs to consist of individuals who have specific key skills and competencies, with the right knowledge and expertise to successfully collaborate with the other functional areas. Some of the key competencies required must be in the fields of:

- Maintenance and reliability
- Operations/business administration
- Project management principles
- Finance and accounting

A physical asset management team consisting of such individuals will be well balanced and will cover the whole spectrum of knowledge areas and expertise required for a holistic physical asset management philosophy.

Physical Asset Knowledge Base

A *physical asset knowledge base* is a key requirement for an effective holistic physical asset management implementation. We have seen, in the earlier chapters, that all functional areas are important in their own ways to the organization. All must coexist and work together in harmony to perform their duties: one function is tasked to perform certain activities on the physical asset; others will be carrying out planning functions; and, another may be focused on making investment decisions. In the real estate world, we know the mantra is location, location, and location and, in physical asset management, it is decision, decision, and decision. To make the right decisions, the physical asset management function, especially in physical-asset-intensive organizations, relies heavily on a solid physical asset knowledge base. Typically, the physical asset knowledge base will consist of physical asset data (technical, financial and non-financial), information, and knowledge. The knowledge base can be in the form of physical asset inventory, classification, description, hierarchy, technical specification, financial data, supplier and manufacturer names, drawings, operation and maintenance manuals, performance standards, conditions, failure modes, maintenance procedures, spare parts inventory, maintenance and repair history, maintenance and repair cost, performance indicators, and many other relevant features.

As we see, the physical asset knowledge base originates from the physical asset itself, from the activities performed on the physical asset, and from the experience of the physical asset users and owners. All must be accurate, up-to-date, accessible, understood, and managed in order to ensure that the physical asset knowledge base provides effective support to the optimized decision making processes.

"I am sure most, if not all, organizations would have that," you say.

I am not so sure about that. This is one of those areas in physical asset management that really needs to be improved. Typically, organizations would struggle to capture accurate physical asset data, information, and knowledge on a consistent basis, resulting in a low-quality, substandard physical asset knowledge base that is not always reliable for good decision making. The bottom line is that, for effective implementation of holistic physical asset management, organizations must have their physical asset knowledge base well organized, up-to-date, and to a certain standard, to enable effective use of it. A physical asset management information system is the solution to this problem.

Physical Asset Management Information System

A physical asset management information system is crucial for a holistic physical asset management approach because it links up all the physical assets' data and make them useful information ready to be used in the optimal decision making process. As per the International Infrastructure Management Manual, a ***physical asset management information system*** is defined as: "a combination of processes, data, software and hardware applied to provide the essential outputs for effective asset management such as reduced risk and optimum infrastructure investment."

In this information age, it is unthinkable to picture a physical asset intensive organization that deals with lots of physical asset data to be without an asset management information system. It has become a necessity for many organizations to collect, store, process, and analyze asset data and information to better manage their physical assets over their whole lifecycle. An asset management information system can be as complex as an Enterprise Asset Management software (EAM) with different functionalities, or some customized spreadsheets for data analysis, depending on the size, complexity of the organization and its requirements.

At the end, the physical asset management function would want an information system to include, at minimum, the following:
- Up-to-date physical asset registry and inventory
- Detailed physical asset maintenance program
- Work management system to plan and record work
- Physical asset condition assessment and monitoring
- Spare parts inventory management
- Physical asset failure and risk analysis
- Physical asset history and lifecycle cost
- Predictive modeling and financial planning

In many organizations nowadays you will find Computerized Maintenance Management Systems (CMMS), Work Management Systems (WMS), Asset Management Information System (AMIS), other databases, and several spreadsheets in use. A lot of times, you will see that either these systems are not up-to-date, are not fully utilized, or are utilized but in silos. The result is that you will never get accurate physical asset data for proper decision-making, or the physical data will be spread out in different systems/databases, making it almost useless. For an effective holistic physical asset management implementation, two key things need to happen with an asset management information system:

1. All the different information systems and databases must be consolidated into one, or integrated with each other so that they are all aligned and consistent to provide useful information.
2. Data must be collected and entered consistently, accurately, and correctly. Truncated, missing, or inaccurate data is of no use in physical asset management because it does not provide the full picture of the physical asset's situation.

CONCLUSION

Organizations have to realize that implementing a holistic approach to manage physical assets is of utmost importance. The truth is that physical asset management is recognized globally and in most industries, as a must to ensure we leave a legacy for future generations. Previous generations have done a great job of building highly performing physical assets to make our lives easier. They were the visionary Baby Boomers. We have to admit that the Generation X has so far neglected to take care of those physical assets as they should be. But it is never too late. Even though we are facing huge backlog, it is our job now to build on what we have inherited, with objective to make the lives of Millennials and future generations better. As newer and newer physical assets are going up, we need to save more today for future asset management needs. Of course we have to accept that it won't come easy due to some worrying existing challenges, such as limited funding, the global economy, climate change, energy preservation pressures, and global trends in general. All this make organizations, business leaders, and physical asset management practitioners vigilant and mindful to manage physical assets with a different approach over their entire lifecycles.

- Organizations must be mindful that the different approach should not be limited to creating asset management plans only, or developing fancy capital plans, or implementing onerous reliability programs. A holistic physical asset management approach is much more than that. It involves whole organizational transformation from top to bottom, total leadership commitment, major organizational behaviour and culture change and entire workforce engagement. In a nutshell successfully implementing a holistic physical asset management philosophy will require total synergy and alignment across the whole organization and its functional areas. A right governance model spanning across the whole organization will be of great help. Governance plays an important role in determining how organizations function and defines the processes, structures and organizational traditions that determine the leadership and how power is exercised, how stakeholders have their say, how decisions are taken and how decision-makers are held accountable. In a successful governance model, the governing body will be tasked to:

- Create a vision
- Secure resources
- Define clear roles and responsibilities
- Establish benchmarks for performance
- Establish accountabilities to key stakeholders

A governance model suitable for the organization coupled with a well-crafted physical asset management framework will provide the right leadership to focus on 'why adopt a holistic physical asset management philosophy' and 'what is it' with respect to the interactions and collaboration of the different functional areas. This will build a solid foundation for the development of physical asset management policy and strategy at the organizational level. The governance model does not specifically focus or address the 'how' to implement physical asset management as that is more of a tactical decision at the program and physical asset level, plus a recognition that there is no universal approach that meets all needs for different asset classes.

The governance model will confirm the holistic nature of physical asset management in the context of a broad scope of multi-disciplinary and integrated scale of activities to achieve the organizational strategic objectives. All stakeholders are encouraged to assess their role and mandate, in the context of the governance model, recognize how they contribute to the overall objective of the organization and of the physical asset management approach. Through the model, the whole organization and its functional areas can be brought together and made to work in harmony. It will take the right type of governance, strong leadership, the collective and integrated efforts of the whole organization and personnel from different functional areas to address the significant physical asset management challenges.

Good luck on the journey to rediscover the holistic physical asset management philosophy in your organization!

BIBLIOGRAPHY

AMBOK Framework, Asset Management Council ISBN 978-0-9870602-3-5

AMP Capital. Understanding Infrastructure – a reference guide. http://www.ampcapital.com/ampcapitalglobal/media/contents/campaign/real/understanding_infrastructure_a_reference_guide.pdf.

An Asset Management Governance Framework for Canada. NAMWG-GNTGA, 2009. https://www.engineerscanada.ca/public-policy/national-round-table-on-sustainable-infrastructure.

Asset Management Council. Information Guide. Engineers Australia. https://www.amcouncil.com.au/files/Asset_Management_Council_1506_Information_Guide.pdf.

Asset Management Council. The Asset Management Concept Model. The Asset Journal, Issue 4, Vol. 7, 2013. E-Copy.

Asset Management Landscape Second Edition, ISBN978_0_9871799_2_0_ GFMAM Landscape, Second Edition. E-Copy

Asset Management: The Future of Business Management. The Asset Journal, 2013. E-Copy.

Botha, Arnold. Is ISO 55000 An Oxymorom, Or Merely The Inner Circle of Asset Management?. Pragma. http://www.pragmaworld.net/pdfs/commentaries/Is%20ISO%2055000%20an%20oxymoron,%20or%20....pdf.

Boudreaux, Kenneth J. Finance. Edinburgh Business School, Heriot-Watt University. 2003. Print.

Campbell, John D. and James V. Reyes-Picknell. Uptime. 2nd ed. New York City: Productivity Press, 2006. Print.

Campbell, John D., Andrew K. S. Jardine and Joel McGlynn. *Asset Management Excellence*. 2nd ed. Boca Raton: CRC Press, 2011. Print.

Canada Infrastructure. *Canadian Infrastructure Report Card: Asset Management Primer*. Canada Infrastructure. http://www.canadainfrastructure.ca/downloads/circ_asset_management_primer_EN.pdf.

Canada Infrastructure. *Canadian Infrastructure Report Card: Informing the Future*. Canada Infrastructure. http://www.pppcouncil.ca/web/News_Media/2016/2016_Canadian_Infrastructure_Report_Card__Informing_the_Future.aspx.

Dailey, Robert. *Organizational Behaviour*. Edinburgh Business School, Heriot-Watt University. 2012. Print.

Daley, Daniel T. *Design For Reliability*. New York City: Industrial Press Inc., 2011. Print.

Davis, Mark M., Janelle Heineke, and Jaydeep Balakrishnan. *Fundamentals of Operations Management*. 2nd ed. Boston: McGraw-Hill Ryerson Ltd., 2007. Print.

Davis, Robert. *An Introduction to Asset Management*. The Insititute of Asset Management. E-Copy.

Distefano Bob. Goetz, Will. Storino, Bruce. *Operational Readiness: Bridging the Gap Between Construction and Operations for New Capital Assets*. People and Processes, 2015. http://reliabilityweb.com/articles/entry/bridging_the_gap_between_construction_and_operations/.

Federation of Canadian Municipalities. *Decision Making and Investment Planning: Selecting a Professional Consultant*. National Guide to Sustainable Municipal -Infrastructure, 2006. https://www.fcm.ca/Documents/reports/Infraguide/Selecting_a_Professional_Consultant_EN.pdf.

Freitas, Italo. *Striving For Excellence: Implementation of the Asset Management System*. The Asset Journal, Issue 3, Vol. 8. E-Copy.

Goldsmith, Marshall and Mark Reiter. *What Got You Here Won't Get You There*. New York City: Hyperion, 2007. Print.

Guide to Accounting For and Reporting Tangible Capital Assets. 2007. http://www.frascanada.ca/standards-for-public-sector-entities/resources/reference-materials/item14603.pdf.

Gulati, Ramesh. *Maintenance and Reliability Best Practices*. New York City: Industrial Press Inc., 2009. Print.

Haarman, Mark, and Guy Delahay. *Value Driven Maintenance – New faith in maintenance*. KJ Dordrecht: Mainnovation, 2004. Print.

Hansen, Robert C. *Overall Equipment Effectiveness*. New York City: Industrial Press Inc., 2001. Print.

Hastings, Nicholas A.J. *Physical Asset Management*. London: Springer-Verlag, 2010. Print.

https://en.wikipedia.org/wiki/Asset

https://en.wikipedia.org/wiki/Asset_management

https://en.wikipedia.org/wiki/Infrastructure

https://en.wikipedia.org/wiki/Infrastructure_asset_management

IBM. *Enabling The Benefits of PAS 55: The New Standard For Asset Management In The Industry*. White Paper, 2009. https://www-935.ibm.com/services/uk/bcs/pdf/090622_pas_55_white_paper_6815_tiw14035usen00_final.pdf.

Ilao, Toribio. *Using Physical Asset Management As A Strategy For Cultural Transformation*. The Asset Journal, Issue 4, Vol. 8. E-Copy.

Infrastructure Asset Management Exchange. *From Inspiration To Practical Application: Achieving Holistic Asset Management*. The Institute of Asset Management, 2015. E-Copy.

Infrastructure Report Card. *Report Card for America's Infrastructure*. Infrastructure Report Card, 2013. http://www.infrastructurereportcard.org/.

International Infrastructure Management Manual. IPWEA (ed.), Australia. 2006.

ISO 31000:2009 Risk management - Principles and guideline;

ISO 31010:2009 Risk management - Risk assessment techniques

ISO 55000: Asset Management - Overview, principles and terminology. ISO 2013

ISO 55001: Asset Management - Management systems - Requirements. ISO 2013

ISO 55002: Asset Management - Management systems - Guidelines for the application of ISO 55001. ISO 2013

Lafraia, JR and J Hardwick. Living Asset Management. Crows Nest: Engineers Media, 2013. Print.

Lloyd, Chris. Asset Management – Whole-life management of physical assets. 2nd ed. Westminster: ICE Publishing, 2010. Print.

Lloyd, Chris. International Case Studies in Asset Management. London: ICE Publishing, 2012. Print.

Lockyer, Keith, et al. Production and Operations Management. 5th ed. London: Pitman Publishing, 1988. Print.

Lothian, Niall; Small John. Accounting. Edinburgh Business School, Heriot-Watt University. 2014. Print.

Lutchman, Roopchan. Sustainable Asset Management. Lancaster: DEStech Publications, Inc., 2006. Print.

Mackenzie, Hugh. Canada's Infrastructure Gap: Where It Came From and Why It Will Cost So Much To Close. Canadian Centre for Policy Alternatives, 2013. https://www.policyalternatives.ca/sites/default/files/uploads/publications/National%20Office/2013/01/Canada's%20Infrastructure%20Gap_0.pdf.

McKinsey Global Institute, Bridging Global Infrastructure Gaps, McKinsey & Company, June 2016. http://www.mckinsey.com/industries/infrastructure/our-insights/bridging-global-infrastructure-gaps.

McKinsey Global Institute, Infrastructure Productivity: How to save $1 trillion a year, McKinsey & Company, January 2013. http://www.mckinsey.com/industries/infrastructure/our-insights/infrastructure-productivity.

MIL-HDBK-470A, August 1997, Department of Defence Handbook Designing and developing maintainable products and systems, Volume 1

MIL-STD-973 Military Standard Configuration Management

Mind the Gap: Finding the Money to Upgrade Canada's Aging Public Infrastructure. TD Bank Financial Group, 2004. https://www.td.com/document/PDF/economics/special/td-economics-special-infra04-exec.pdf.

Mirza, Saeed. *Danger Ahead: The Coming Collapse of Canada's Municipal Infrastructure*. Federation of Canadian Municipalities, 2007. https://www.fcm.ca/Documents/reports/Danger_Ahead_The_coming_collapse_of_Canadas_municipal_infrastructure_EN.pdf.

Mitchell, John S. *Physical Asset Management Handbook*. 4th ed. Fort Myers: Reliabilityweb.com, 2013. Print.

Morgan, Jacob. *The 12 Habits of Highly Collaborative Organizations*. Chess Media Group. http://www.forbes.com/sites/jacobmorgan/2013/07/30/the-12-habits-of-highly-collaborative-organizations/#258a91985f12.

Moubray, John. *Reliability-centered Maintenance*. 2nd ed. Woodbine, NJ: Industrial Inc., 1997. Print.

Mullins, John; Walker, Orville C Jr; Boyd, Harper W Jr; Jamieson, Barbara. *Marketing*. Edinburgh Business School, Heriot-Watt University. 2010. Print.

Narayan, Vee. *Effective Maintenance Management Risk and Reliability Strategies for Optimizing Performance*. New York City: Industrial Press Inc., 2004. Print

National Research Council Canada. *MIIP Report: A Primer on Municipal Infrastructure Asset Management*. Institute for Research in Constitution, 2004. http://nparc.cisti-icist.nrc-cnrc.gc.ca/eng/view/fulltext/?id=bf6708c0-24ea-4c7f-8c85-f8d45c87399d.

O'Connor, Patrick D. T., et al. *Practical Reliability Engineering*. 4th ed. Chichester: John Wiley & Son Ltd., 2002. Print.

Penrose, Howard W. *Physical Asset Management for the Executive*. Old Saybrook: SUCCESS by DESIGN Publishing, 2008. Print.

Peterson, Bradley. *The Future of Asset Management*. The Asset Journal, 2013. E-Copy.

Procurement Strategies. UNSW – Contracts Management. E-Copy.

Procurement. https://en.wikipedia.org/wiki/Procurement

Product Assurance: Reliability, Availability, and Maintainability. Headquarters Department of the Army, 2015. E-Copy.

Project Management Institute, *A Guide to the Project Management Body of Knowledge (PMBoK Guide)*, Fifth Edition, 2013. E-copy.

RealityWeb.com: Developing an Asset Management Strategy. 2015. http://reliabilityweb.com/articles/entry/developing_an_asset_management_strategy.

ReNew Ontario – A five-year infrastructure investment plan to strengthen our economy and communities. Queen's Printer, 2005. http://www.infrastructure.gc.ca/prog/agreements-ententes/bcf-fcc/on-eng.html.

Scott, Alex. Strategic Planning. Edinburgh Business School, Heriot-Watt University. 2007. Print.

Smith, Thomas. Changing Definitions of Asset Management: The Impact of ISO 55000. The Asset Journal, Issue 3, Vol. 8. E-Copy.

Tett, Gillian. The Silo Effect. New York City: Simon & Schuster, 2015. Print.

The City of Calgary – TCA Program Office. TCA Best Practices Benchmarking Survey Findings Report. Canadian Network of Asset Managers, 2011. http://www.cnam.ca/uploads/File/Association%202011/TCA_Best_Practices_Benchmarking_Survey_Results_Nov_2011.pdf.

The Institute of Asset Management, An Anatomy of Asset Management, Version 3, December 2015. E-copy.

The Management and Control of Quality. South-Western/Thomson Learning, 2002. E-Copy.

Thomson, Alex. The SALVO Project: Innovative Approaches to Decision-Making For The Management of Aging Physical-Assets. The Woodhouse Partnership Ltd, 2011. http://www.twpl.com/twpl-library/asset-management-white-papers/salvo-project-innovative-approaches-decision-making-management-aging-physical-assets/.

UBS. An Introduction to Infrastructure As An Asset Class. UBS Global Asset Management. E-Copy.

Uddin, Waheed, W. Ronald Hudson, and Ralph Haas. Public Infrastructure Asset Management. 2nd ed. New York City: McGraw-Hill Education, 2013. Print.

Woodhouse, John. Asset Management: Concepts & Practices. The Woodhouse Partnership Ltd, 2006. http://reliabilityweb.com/articles/entry/asset_management_concepts_practices/.

Woodhouse, John. ISO 55000 – What, Why and How: An Introduction To The First International Standard For Asset Management. The

Woodhouse Partnership Ltd, 2014. http://vision-aeam.com/wp-content/uploads/2014/04/3-1-Brief-intro-to-ISO55000.pdf.

Woodhouse, John. *Managing Mature Assets*. The Woodhouse Partnership Ltd, 2009. E-Copy.

Woodhouse, John. *Optimal Timing for Replacing Aging or Obsolete Assets*. The Woodhouse Partnership Ltd, 2011. http://www.twpl.com/twpl-library/asset-management-white-papers/optimal-timing-replacing-aging-obsolete-assets/.

INDEX

Accounting, 25, 58, 62, 138
Acquisition, 18, 63, 73, 108, 113, 122, 124, 128, 131
Acquisition process, 129
Allocation of authority, 36
Areas of focus, 41, 43, 44, 59, 64, 119
Asset, 4, 5, 6, 7, 24, 26, 27, 28. See also Physical assets
Asset lifecycle costs, 17
Asset Management Information System (AMIS), 180
Asset Management Landscape, 51, 174, 175, 184
Asset Management System, 53, 171, 186
Asset manager, 176, 177
Asset planning, 132
Association of Asset Management Professionals (AMP), 19
Availability, 77, 82, 95, 103, 104, 106, 107
Backlog, 135, 154, 181
Bottom-up approach, 19, 45, 46, 56, 152
Build-own-operate-transfer, 125
Capacity, 78, 83, 93, 95, 111, 117, 139
Capital investment, 130, 134
Capital planning, 55, 58, 134, 153, 176
Capital projects, 26, 110, 111, 113, 115, 116, 118, 134, 150
Cash flow, 135

CMMS, 86, 87, 93, 94, 95, 118, 151, 180
Collaboration, 161, 165
Communication, 2, 36, 37, 70, 72, 80, 93, 94, 123, 152, 156, 160
Condition assessment, 140, 146
Configuration management, 112, 129, 149
Consequences, 2, 18, 84, 88, 89, 99, 100, 147
Construction management, 125
Corporate strategy, 21
Corrective, 89
Cost, 14, 25, 32, 43, 63, 64, 81, 90, 91, 94, 95, 111, 113, 115, 120, 128, 160, 161, 162
Cost benefit analysis, 95
Cost optimization, 48, 91, 165
Cottage industry, 9, 11, 15
Craft production era, 9
Criticality, 28, 100, 101
Culture, 37, 40
Data management era, 9
Decision making, 17, 30, 32, 36, 41, 66, 140, 147, 162, 173, 178
Demand, 14, 23, 76, 79, 93, 145, 146, 173
Demand management, 145, 146, 173
Departmentalization, 36
Depreciation, 139
Design modification, 93

Design-construct-maintain, 125
Design-develop-construct, 125
Detailed design and construct, 125
Detective, 86, 89
Discreet physical assets, 29
Disposal, 18
Economic life, 18, 22, 67, 80, 84, 85, 86, 96, 128, 139, 144, 146
Effects, 2, 99
Engineering, 18, 26, 61, 62, 97, 150, 191
Enterprise Asset Management software (EAM), 179
Equipment, 4, 15, 23, 25, 29, 30, 32, 68, 117
Equipment-Centric Organizations, 30
Evaluate, 14, 99, 144
Evolution, 9, 11, 13, 14, 16, 19, 33, 109
Excellence, 7, 12, 36, 85, 89, 160, 161
External agencies/outside contractors, 115
Facilities, 16, 29, 30, 32, 133, 134
Factory system, 10, 15
Failure modes, 178
Failure modes and effects analysis (FMEA), 99
Failure rate, 104
Failure reporting analysis and corrective action system (FRACAS), 100
Finance, 7, 23, 25, 64, 74, 83, 115, 153, 158, 167, 176
Financial assets, 27
Financial planning, 134, 135, 136, 140, 153, 155, 180
Financial reporting, 139, 140
Financial statement, 138, 141
Financial strategy, 130, 138, 140
Financial sustainability, 77
Five Ps, 75
Fixed, 15, 66, 74, 81, 140
Forecast, 13, 95, 134, 135, 168
Function, 6, 7, 12, 19, 23, 24, 25, 56, 57,

Functional areas, 7, 20, 23, 26, 32, 34, 37, 41- 46, 59, 62, 64- 67, 72-74, 79, 82, 83, 101, 142, 144, 155, 156, 157, 167, 168, 169, 174, 177, 182
Functional silos, 44, 45, 59, 60, 66, 70, 71, 72, 167
Functional structure, 109
Gaps, 3, 7, 44, 47, 56, 72-74, 93, 102, 115, 148, 156, 160
Generally accepted accounting principles (GAAP), 138
Global Forum of Maintenance and Asset Management (GFMAM), 19
Governance, 164, 182, 184
Governance model, 164, 182
Governmental Accounting Standards Board, GASB 34, 138
Harmonized, 45, 71, 72, 79, 171
Harmonized interactions, 148
Harmonized silos, 71
High-intensity maintenance, 68, 69, 136
Holistic approach, 5, 17, 18, 32, 55, 60, 114, 143, 144, 148, 181
Holistic physical asset management, 5, 20, 34, 35, 48, 55, 66, 142, 144, 156, 161, 167, 177, 182
Holistic Physical Asset Management House of Excellence, 161
Human assets, 27
Human resources, 62
Industrial revolution era, 9
Information assets, 27
Information database, 86
Information technology, 13, 23, 26, 29
Infrastructure, 2-4, 17, 23, 27, 29, 30, 32, 58, 159, 179
Infrastructure needs, 3, 4
Infrastructure productivity, 3, 115
Infrastructure-Centric Organizations, 31, 68, 136
Inherent reliability, 95, 96, 103, 107, See also Reliability

Input, 81, 122
Installation/commissioning, 63, 73
Institute of Public Works Engineering Australasia (IPWEA), 18
Intangible assets, 27
Interaction, 23, 27, 37, 40, 47, 62, 66, 72, 73, 79, 148, 158, 160
International Financial Reporting Standards (IFRS), 138
International Infrastructure Management Manual (IIMM), 18, 49, 60
Inventory management, 179
Investment backlog, 154
Investment gaps, 4
Investment needs, 3, 132, 134
ISO 31000 Risk Management Principles and Guidance, 147, 188
ISO 55000, 5, 7, 18, 27, 49, 50, 51, 53, 60, 171, 172
Leadership, 22, 37, 38, 40, 54, 166, 169, 172, 176, 182, 183
Lean production era, 9, 10, 13
Level of service, 63, 78, 147, 165, 173
Lifecycle costs, 128
Lifecycle delivery activities, 7, 56, 68, 69, 84, 93, 94, 102, 167, 176
Line of sight, 45, 46, 47, 59, 72, 100, 102
Line of Sight, 45, 47
Linear physical assets, 29
Long-term needs, 133, 167
Long-term sustainability, 169
Low-intensity maintenance, 67, 68, 69, 137
Main functions, 23, 75, 169
Maintenance, 5, 15, 16, 18, 19, 25, 31, 43, 50, 55, 56, 58, 63, 67, 73, 77, 80, 85-96, 97-98, 100, 102, 103, 104, 115, 117, 118, 128, 137, 151, 152, 154, 155, 156, 157, 162, 167, 168, 178
Maintenance efficiency, 90

Maintenance management, 44, 55, 56, 57, 88, 89, 91-93, 95, 96, 107
Maintenance strategies, 89
Maintenance tactics, 89, 96, 118
Management by objective (MBO), 41
Marketing, 25, 62, 191
Matrix structure, 109
Mean time between failures (MTBF), 104
Mechanistic, 36, 37, 38
Mission statement, 21
Mean time to repair (MTTR), 106
Needs, 3, 14, 18, 23, 64, 83, 84, 109, 111, 114, 119, 122, 124, 126, 130, 133, 134-136, 145, 149, 153, 156, 162, 167
Needs assessment, 145
OCS Approach, 161, 162, 165, 169
Operating context, 28, 52, 98, 101, 102, 104, 106, 107
Operational effectiveness, 36
Operational needs, 82
Operations, 9, 10, 18, 23, 25, 44, 58, 64, 73, 75-80, 84, 86, 92, 95, 103, 116, 117, 131, 157, 162
Operations function, 23, 25, 45, 76, 79, 82, 83, 84, 92
Optimum costs, 77
Optimum decision, 37, 143, 144, 147, 162
Organic, 36, 37
Organizational design, 36
Organizational level, 54, 69, 182
Organizational silos, 41, 59, 71
Organizational strategic goals, 22, 36, 37
Organizational structure, 37
Organizations, 4, 5, 6, 7, 8, 9, 14, 19, 21, 28, 30, 31, 32, 34-39, 41, 45-47, 54, 59, 60, 69, 81, 130, 136, 142, 160, 161, 172, 181
Output, 15, 41, 45, 76, 81, 82, 91, 162

Overall equipment effectiveness (OEE), 82
Overlaps (G&O), 45, 73
Ownership, 161, 162
Paradigm shifts, 8
PAS 55, 7, 18, 49, 50, 171, 187
PAS55, 60
People, 8, 18, 22, 23, 26, 28, 36, 38, 39, 40, 53, 60, 72, 170
Performance, 8, 15, 26, 32, 40, 43, 45, 49, 78, 81, 84, 94, 95, 102, 128, 147, 172, 173, 176, 178, 182, See also Performance standards
Performance efficiency, 82
Performance indicators, 95, 178
Performance standards, 26, 94, 95, 100, 109, 118, 128, 150
Physical asset data, 178
Physical asset failures, 31, 44, 88
Physical asset knowledge base, 177
Physical asset level, 66, 67, 182
Physical asset lifecycle phases, 17, 18, 63, 64, 65, 73, 108, 144
Physical asset management, 1, 5, 6, 7, 18, 19, 20, 32, 33, 34, 35, 47, 48, 49, 54-60, 66-70, 72, 73, 74, 142-144, 146, 149, 155, 156, 159, 160-164, 167, 168, 169, 170, 172, 174, 176, 177, 178
Physical asset management collaborative efforts, 168
Physical asset management framework, 172
Physical asset management function, 19, 32, 33, 36, 40, 47, 53, 55, 57, 58, 74, 142, 144, 148, 150, 153, 158, 164-167, 168, 169
Physical asset management information system, 179
Physical Asset Management Structure, 176
Physical assets, 3, 23, 28, 29, 31, 32, 46, 67-69, 76-78, 86, 88, 89, 101, 102, 133, 136
Planetary gear system, 157, 158
Planned maintenance completion ratio, 96
Planning, 18, 51, 130, 132, 133, 144, 145, 154, 172, 186, 192
Plant, 21, 22, 23, 25, 28, 61, 77, 80, 92, 93, 116, 117, 126, 134
Plant Engineering Maintenance Association of Canada (PEMAC), 18
Predictive, 89
Preventive, 15, 86, 89
Private sector, 7, 19, 29, 35, 59, 68, 70
Proactive maintenance, 103, 154
Procurement, 18, 25, 43, 58, 62, 64, 83, 86, 121-129, 157, 168
Procurement process, 121, 122, 124, 126, 128
Product, 2, 11, 12, 13, 14, 22, 40, 58, 100, 116, 117, 134
Productivity, 81
Project benefits, 109, 115, 119
Project management, 26, 57, 62, 73, 108-115, 126, 149, 152, 162
Project Management Body of Knowledge (PMBoK), 110
Project management function, 112, 113, 126, 149
Project management/engineering, 82
Project objectives, 109, 110, 115, 119
Projectized structure, 109
Public sector, 7, 17, 29, 57, 58, 59, 67, 123, 131, 136, 138, 139, 159
Pull angle, 14
Push angle, 14
Quality, 11-15, 23, 24, 25, 27, 43, 48, 82, 83, 122, 124, 128
Quality management, 10, 12
Quality rate, 82

RCM methodology, 85, 98, 99, See also Reliability-centered maintenance
Refurbishment, 32, 79, 93, 111, 155
Rehabilitation, 3, 63, 67, 134, 136, 137, 139, 154
Reliability, 14, 16, 19, 20, 30, 87, 95, 96, 97-107, 116, 118, 149, 151, See also Inherent reliability
Reliability engineering, 97, 106
Reliability-centered maintenance, 16, 77, 98
Reliabilityweb, 19, 190
Replacement, 31, 32, 79, 93, 111, 134, 136, 137, 139, 147, 148, 154
Ring gear, 157, 158
Risk, 2, 18, 32, 33, 48, 49, 91, 99, 111, 113, 128, 145, 147, 148, 156, 159, 160, 161, 162, 165, 167, 173, 176, 179, 180
Risk assessment, 146, 147, 188
Return on investment (ROI), 139, 140
Root cause failure analysis (RCFA), 100
Service level, 17, 32, 46, 78, 83, 100-102, 115, 134, 147, 148, 152, 161, 162
Service level assessment, 146, 147
Social responsibility, 77
Span of control, 36
Stanley Nowlan and Howard Heap, 16
State of good repair (SOGR), 133
Strategic goals, 22, 23, 36, 46, 63, 100, 109, 173
Strategic objectives, 18, 21, 26, 27, 32, 39, 46, 51, 75, 85, 136, 148, 152, 172, 173, 176, 182
Strategic thinking, 143
Supporting functions, 75
Sustainment, 161, 169
Synergies, 7, 66, 93, 148
Synthesize, 143, 144
System engineering, 112, 149
System reliability, 100, 104, 105, 107

Tangible assets, 4, 27, 138
The Asset Management Council (AMC), 18
The Institute of Asset Management (IAM), 18
Three Ps, 21, 22, 23, 28
Time, 44, 109, 110, 119, 153
Total quality management (TQM), 12, 76
Traditional construction, 125
Transformation Process, 76
Triangle of focus (ToF), 42
Unplanned, 88
Value, 7, 18, 24, 27, 28, 34, 54, 70, 76, 78, 90, 100, 109, 118, 139, 145, 148, 162
Vision statement, 21
Wish list, 134, 153
Work management, 96, 151
Work Management Systems (WMS), 180
Work specialization, 36

Printed in Canada